1. 40

Statistics
in theory and practice

Statistics
in theory and practice

L R Connor
MSc(Econ), FSS
Farr Medallist

and

A J H Morrell
MA, FIS, FSS, AIA
Late Scholar of Trinity College, Cambridge

Sixth edition by H A J Morrell

Pitman Publishing

Sixth edition 1972

Sir Isaac Pitman and Sons Ltd
Pitman House, Parker Street, Kingsway, London WC2B 5PB
PO Box 46038, Portal Street, Nairobi, Kenya

Sir Isaac Pitman (Aust) Pty Ltd
Pitman House, Bouverie Street, Carlton, Victoria 3053, Australia

Pitman Publishing Company SA Ltd
PO Box 11231, Johannesburg, South Africa

Pitman Publishing Corporation
6 East 43rd Street, New York, NY 10017, USA

Sir Isaac Pitman (Canada) Ltd
495 Wellington Street West, Toronto 135, Canada

The Copp Clark Publishing Company
517 Wellington Street West, Toronto 135, Canada

Cased edition ISBN: 0 273 36160 0
Paperback edition ISBN: 0 273 36155 4

Set in 10/11 pt. Monotype Times New Roman, printed by
letterpress, and bound in Great Britain at The Pitman Press, Bath
G2(B2154/1047:43)

Preface to Sixth Edition

Many years ago a disillusioned man said, "Of making many books there is no end." He might well have been writing of the spate of books on statistics that have been published in recent years, often covering well-trodden ground. A new edition of an existing book can, of course, be justified by the plea, "We got in first." In any case, however, I believe this book is unique in providing not only an introduction to statistical methods but a fairly comprehensive and (at the time of writing) up-to-date account of British published statistics.

At the present time there are some interesting developments in official statistics and, partly because the scene is constantly changing, a certain amount of detail has been omitted from the later chapters. Also I thought it desirable to make it less like a reference book. Some of the earlier chapters have been expanded and (I hope) improved, notably those dealing with surveys, time series, sampling and index numbers. I am greatly indebted to Mr D. Walley for revising and expanding Appendix 1 on Data-Processing Techniques and to International Computers Limited for providing new photographs for it.

This book is intended for students with a leaning towards business and economics, and in particular for those studying for the examinations of professional bodies like the Institute of Cost and Works Accountants, the Association of Certified and Corporate Accountants and the Institute of Transport. It should be specially useful to students taking Economic and Social Statistics as their special subject in Part II of the Institute of Statisticians' examinations.

I wish to thank the four bodies named above for permission to reproduce their examination questions, the Controller of H.M.

Stationery Office for permission to reproduce copyright matter and Mrs Josephine Parkinson for so efficiently typing the new material. I should also like to record my great debt to the original author of this book, the late L. R. Connor. Although little of the original material remains, the concept and general plan of the book are still his.

A J H Morrell

February 1972

Abbreviations

The following abbreviations are used to indicate the sources of examination questions reproduced as exercises at the end of suitable chapters—

ACCA	*Association of Certified and Corporate Accountants*
ICWA	*Institute of Cost and Works Accountants*
IoS	*Institute of Statisticians*
IoT	*Institute of Transport*

Contents

APPENDIXES

List of Tables

List of Figures

List of Plates

1

Introductory

THE word "Statistics" originally referred to collections of facts, not necessarily numerical, about the State, or the people who composed the State. According to modern usage it refers to collections of numerical facts or estimates and any restriction to affairs of state is obsolete. It may be construed either as singular or as plural. Statistics (plural) are the figures themselves, suitably classified and tabulated, together with any secondary statistics such as percentages or averages derived from them; it is in this sense that the public usually think of statistics. Statistics (singular) is the study, better described as Statistical Methods, which deals with the collection, analysis and interpretation of the figures.

Statistical Methods may be broadly divided into Descriptive Statistics and Mathematical Statistics. Descriptive Statistics deals with the compilation and presentation of data as actually recorded, generally in the course of public or private business, not for the purpose of refined analysis but simply in order to provide concise information on which decisions can be taken. Mathematical Statistics is based on the theory of probability and attempts to draw precise general conclusions from the data. It may also help to decide how the data can be obtained most efficiently and economically. In recent years there have been great advances in the design of experiments in biology, agriculture and industry with the aid of statistical methods. In economic and commercial problems, however, it is difficult to conduct experiments, and the scope for refined methods and mathematical techniques is severely limited.

Just as there are different kinds of statistics, so there are different kinds of statisticians. Indeed, the subject of statistics is now so wide that it is impossible for one man to become expert in all branches.

The industrial statistician may know little about economic statistics, and the economic statistician may know little about the mathematical techniques used in scientific and industrial research. The present tendency, however, is to require some knowledge of the whole field including operational research and the elements of computers.

The Purpose of Statistics

It cannot be too strongly stressed that the purpose of statistics is not simply to satisfy curiosity but to provide estimates or comparisons on which sound decisions can be made. Thus, if a manager is told that Factory A produced 500 tons of a certain product in a week, he must have some standard of comparison to tell him whether that is satisfactory. He may judge that the capacity of the factory is 650 tons a week; or he may be given a series of figures showing that it normally produces about 600 tons a week; or again, he may be told that Factory B, with half as many workers but otherwise comparable with Factory A, produced 350 tons in the same week. Such comparisons provide information on which action can be taken.

Looking further ahead, estimates based on present knowledge are essential to government and industry alike when they are formulating plans for the future. A local authority planning to build new schools must estimate the numbers of children likely to require admission during the next few years. A manufacturing firm deciding whether to build a new plant, or launch an export drive or an advertising campaign, requires estimates of future demand as well as knowledge of current trade and production statistics. Such estimates may be based partly on statistics of population, incomes, regional employment, etc., and partly on a mathematical analysis of past records projecting into the future. They must, of course, be used with discretion, as forecasts may be falsified by political events, economic crises, new inventions and other unforeseen circumstances; nevertheless, they must be made, and as wisely as possible.

It has been said that every statistical table or chart should be preceded by the question "What for?" and followed by the question "So what?" This advice need not be taken too literally, but the idea behind it is sound. There is often a temptation to compile statistics for their own sake, which must be sternly resisted. *The purpose of statistics is to enable the recipients to take decisions.* The decision may be negative: that does not matter. If the statistics show that no action is required, that does not mean that they were wasted: they may save time that would otherwise be spent in argument.

It must always be remembered that statistics deal only with measurable aspects of things and therefore can seldom give the complete solution to a problem. They provide a basis for judgment but not the whole judgment. Moreover, statistics can be, and often are, used, consciously or otherwise, to justify decisions reached on other grounds. It is therefore important that the limitations of statistics should be understood both by those who compile them and by those who use them. This warning will be repeated from time to time and in various forms throughout this book.

Statistical Societies

A word about statistical societies in the United Kingdom may not be out of place. There are two main bodies, the Royal Statistical Society and the much younger Institute of Statisticians. As their titles suggest, the former is primarily a learned society, while the latter is a professional institution. The division is in many ways unfortunate, as it means that the profession does not present a united front like, say, the Institute of Actuaries. The Institute of Statisticians holds examinations for the Associateship, Part III being of degree level. Nevertheless, there is a place for those—and there are many—who are capable of passing Part I, or I and II, but not Part III. They are entitled to become Registered Statistical Assistants (on passing Part I) or Senior Statistical Assistants (on passing Part II).

2

Statistical Data

Formulating the Problem

Before initiating a statistical return or undertaking a statistical inquiry, the statistician should try to ascertain exactly what is required, and for what purpose. His duty is not only to compile statistics but to interpret them, and even, on occasion, to say that the question cannot be answered by statistics. He should therefore know what use is likely to be made of the results and if necessary, how soon and how accurately they are required. The last two requirements may not be compatible.

It is important to ask the right questions in the first place and then to obtain the right answers. No sophisticated techniques can overcome defects in the basic data. As the computer experts say, "Garbage in, garbage out." The following points should be considered at the outset—

1. Can the required information be given in numerical terms, or can a good substitute for it be found? For example, ability or intelligence cannot be measured directly, but only by results, test scores or examination marks. The incidence of drunkenness can be measured only by numbers of convictions, which will depend partly on the strength or efficiency of the police.

2. What is the precise definition of the object to be measured? In a wage inquiry, for instance, should the data be wage rates or actual earnings, should allowance be made for overtime and bonuses, receipts in kind (e.g. free travel for railwork workers), sick pay and holidays with pay, and should earnings be gross or net after PAYE, superannuation contributions, etc.—always assuming that these alternatives are practicable? (See 1 above.)

3. What should be the field of the inquiry? As will be seen later (Chapter 21), the Census of Production includes extractive and manufacturing industry but excludes transport, distribution and services because of the difficulty of finding suitable units of output. It also excludes establishments employing twenty-five or fewer workers on the grounds that the details obtainable would not be worth the trouble involved in collecting them.

4. Is there any information already available from routine statistics or published sources that will either obviate or supplement further inquiry?

The next question is how to collect the data. Often they can be obtained by direct observation, by counting or measuring. Examples are traffic censuses, times recorded by stop-watch, temperature readings and results of laboratory experiments. In fact, individual items of data are commonly called "observations." Sometimes, however, it is necessary to ask people for the information, either by interview or by some form of postal inquiry. This method will be considered further in Chapter 4.

Data may be divided into Primary and Secondary Data, although authorities differ as to the precise definition of these terms. Broadly speaking, primary data are the raw material, the figures collected at first hand, while secondary data are statistics taken from someone else after being worked up to some extent. The distinction is by no means clear-cut.

Another way is to classify data by source. They may be Internal or External according as they are collected by one's own organization or taken from published statistics or some other outside source. Internal statistics may again be divided into statistical records collected as a matter of routine, and special data obtained by *ad hoc* inquiries to answer specific questions. These two types of internal statistics will be considered in the following two chapters. The latter part of this book deals chiefly with published statistics, but a few remarks may be made at this stage on the nature and interpretation of external statistics.

External Statistics

A wealth of official statistics is published by Government Departments, which naturally possess far greater facilities for collecting reliable information than private industry or individuals and are better able to bear the expense of publication. There are, however, statistics published in technical and trade journals, e.g. *The Rubber Statistical Bulletin*. There are also statistics collected by trade associations and business services, available to members or subscribers. For overseas statistics there are trade returns and other publications issued by foreign governments, the United Nations, and so on. Details of these statistics will be found in Chapters 17 to 24.

Statistics, and especially other people's statistics, are full of pitfalls for the user, and even official statistics must be examined critically. Typical sources of difficulty are—

1. *Changes in Classification*

For example, up to June 1948, numbers registered as unemployed in the United Kingdom relate only to persons insured under the Unemployment Insurance Acts. From July 1948 onwards they include both insured and uninsured persons.

2. *Changes in Geographical and Administrative Areas*

Thus, pre-war figures for Germany have to be compared with post-war figures for Western Germany only, and figures for Great Britain and Northern Ireland must often be compared with earlier figures for Great Britain and the whole of Ireland.

3. *Differences of Definition*

Terms such as "incomes," "profits," "output" and "numbers employed" may be defined in various ways, and this frequently causes confusion. The student may like to try his hand at defining "chemicals" and "chemical industry."

4. *Incompleteness of the Information Provided*

Thus, the annual inquiry by the Department of Employment into earnings and hours of work omits certain important industries such as coalmining and agriculture.

It is safe to say that statistics are hardly ever strictly comparable over a long period of years. The statistician must therefore take pains to ensure that he understands any figures he uses, and is aware of their defects and limitations. Equally, of course, he must be prepared to admit the limitations of his own statistics.

Classification

Classification is the process of arranging things, either in fact or in the mind, in classes each of which possesses some common property or properties relevant to the matter in hand. Thus, in a medical investigation, the incidence of a disease may very well vary between different sexes, different age-groups, different localities and different periods of time. Each of these factors should therefore be recorded for every case of the disease, so that the effect of each one may be evaluated and useful comparisons made between the various classes.

The above illustration exemplifies the main types of classification generally recognized in statistics, which are—

1. *Qualitative*, as when people or things are sorted in groups each possessing some attribute that cannot be expressed numerically. People may, for example, be classified male or female, married or single, or according to their occupation, eye colour, religion, etc.

2. *Quantitative*, as when items vary in respect of some characteristic that can be measured. Such a characteristic is called a *variable* or *variate*. Examples are age, height, income, examination marks, number of children. A variable is said to be continuous when it may pass from one value to the next by indefinitely small steps, e.g. height or age, and discrete when there are gaps between consecutive values, e.g. number of children in a family. In the first case the variable is measured, in the second case it is counted.

The distinction between qualitative and quantitative factors, or attributes and variables, is important because variables lend themselves to mathematical treatment in the form of averages, etc., while attributes do not.

Homogeneity

The finer the classification, the less the chance of including unlike objects under the same head, and the better the chances of being able to compare like with like. For example, it would be misleading to compare the output per employee of two factories if one used more power per head, or more capital equipment per head, than the other. Furthermore, items should be so classified that the totals of the classes make sense. It would be absurd to add 400 motor-cars to 200 prams and call the result 600 vehicles.

It sometimes happens, however, that the variety in the items to be classified is so great that perfect homogeneity within each class can be achieved only by an elaborate and unmanageable classification. Iron and steel production is normally measured in tons, but there is no limit to the variations in shape, size, etc., of iron and steel products, and they differ tremendously in cost per ton and the amount of work required per ton of metal. Non-ferrous metals and alloys are more complicated still. To compress statistical tables for such products into a reasonable compass it is essential to group together items that are not strictly comparable. The same remark applies to articles like drugs, paints and toys. It may even be necessary to express the totals of such items in terms of value only, not of quantity.

Units

A continuous variable can be expressed only in terms of some unit: a distance in terms of yards, metres, miles, etc., a weight in terms of pounds, tons, etc., and so on. These are simple units, but it is often necessary to use composite units such as man-hours, foot-pounds, and feet per second or miles per hour. Some composite units involve three or more simple units. Thus, acceleration can be measured in feet or centimetres per second per second, abbreviated for convenience to ft/sec² or cm/sec². Again, power is measured in foot-pounds per minute or per second.

Care must be taken to ensure that units are, as far as possible, uniform and homogeneous. Thus, many acids and other chemicals are sold in solutions of varying strength, and it would be mis-leading to give their gross weights and add them together. The normal procedure is to express them in terms of 100 per cent acid or some other standard strength. Nitrogenous fertilizers are com-monly expressed in terms of nitrogen content, i.e. as so many tons of nitrogen.

Units should be clearly defined and free from ambiguity. For example, in dealing with American statistics it must be remembered that the US gallon (but not the Canadian gallon) is five-sixths of the Imperial gallon, and the US and Canadian hundredweight means 100 lb, not 112 lb. Tables headed "tons" or "thousands of tons" may refer to long tons (2,240 lb) as in Great Britain, short tons (2,000 lb) as in America, or metric tonnes as on the Continent.

Units must be appropriate for their purpose; e.g., the labour force of an industry should be given in terms of employees, workers, "productive" workers, etc., whichever is most suitable in the context.

Finally, units should be stable. The outstanding example of an unstable unit is a unit of money, since it constantly fluctuates in real value—the trend being usually downward. One remedy is to express values in terms of the £ (say) of some standard year by means of a price index (see Chapter 16).

Derivative Data

A Statistical Derivative is a quantity formed by combining in some way two or more of the original items. The obvious example is a total. This section will be confined to simple derivatives such as ratios and percentages. More complex functions such as measures of dispersion will be considered in later chapters.

There are two ways of comparing two quantities of the same kind, say x and y units. One may be subtracted from, or divided by, the other according as it is the actual or the relative difference that matters. The difference $x-y$ is positive or negative according as x is algebraically greater or less than y, but if the absolute difference is required, it is denoted by $x \sim y$ or $|x-y|$ and called the *modulus* of $x-y$. Thus $|3-5|=2$.

The fraction x/y is often expressed as a ratio $x : y$. Clearly the ratio is unchanged if x and y are both multiplied or divided by the same quantity; e.g., $15 : 5 = 75 : 25 = 3 : 1 = 1 : 1/3$. In words, "15 is to 5 as 75 to 25, or as 3 to 1, etc."

It is frequently necessary to compare two quantities of different kinds. Thus, a town covering 2,450 acres and containing a population of 76,300 persons has a population density of 76,300/2,450, or 31·1 persons per acre. The student may think of many other examples, such as speeds, volumes, densities and coefficients of expansion. Such averages may of course conceal great variations within the town, population, etc.

Percentages must be used with care. Often it is not clear what they refer to, particularly when percentages of percentages are introduced. For example, a certain company was stated to have reduced its dividend "by 100 per cent." Further reading showed, not that the company had "passed" its dividend as might be supposed, but that it had reduced the dividend from 200 per cent to 100 per cent. Such ambiguous statements should be avoided.

Rates

When the numerator of a fraction or the first term of a ratio refers to the number of events or cases occurring during a certain period of time, the fraction or ratio becomes a rate, and is frequently expressed as so many per 1,000. Thus the death rate is the number of deaths per 1,000 of the population during the year in question. The death rate for a quarter is generally expressed as an *equivalent annual rate*, the number of deaths being multiplied by 4 and divided by the population, and similarly for any other period.

Care should be taken that the numerator and denominator are truly comparable, that the denominator represents as accurately as possible the number "exposed to risk," as the actuaries say. A marriage rate, for example, should be calculated on the number of single and widowed persons of marriageable age, and a legitimate birth rate on the number of married women of reproductive age. (See Chapter 18.) Similar considerations apply to averages of the kind considered in the previous section. Thus, the output of a factory can be divided by the total number of employees, the total number of "workers," or the total number of "productive workers," and so on.

When, as generally happens, the number exposed to risk varies over the period, it is customary to take an average. On this basis the accident rate in a factory might be computed by dividing the number of accidents occurring during the year by the average number of workers on 52 weekly payrolls. For a population the mid-year figure is generally taken as the denominator.

Another important point to remember is that an overall rate depends very largely on the structure of the population. Death and sickness rates, for instance, vary greatly according to sex and age, and changes in the crude death rate, covering the whole population, may be due largely to the varying age and sex structure of the population. This point will be further discussed in Chapter 18.

3

Routine Statistics

MUCH has been written both about market research and about published statistics, but the routine statistics compiled regularly for purposes of day-to-day administration have been strangely neglected. This is largely due to the enormous variety of such statistics and of the organizations collecting them, which makes it extremely difficult for any one person to speak with authority on the subject. However, the attempt must be made. It will be convenient to consider the statistics required by a large manufacturing firm, but much of what follows will apply equally to retail firms, local authorities, transport organizations, etc.

Routine statistics are essential to the efficient management of all but the smallest organizations. Many of them, such as those that emanate from the company's accounts, are necessary to comply with the requirements of the law or of Government Departments. Others, such as statistics of sales and costs, are generally needed to ensure that a firm is run profitably, and others, such as statistics of stocks and production, to ensure that it runs smoothly.

Apart from financial and cost statistics, which will probably be left to the accountants, the statistics maintained by an industrial organization can broadly be classified as Sales, Production, Supply (i.e. purchases and stocks) and Personnel. Each of these groups will be dealt with in turn, but certain general considerations applicable to all of them must first be discussed.

Basic Records
Routine statistics are generally derived from (i) permanent records, e.g. of staff or customers, and (ii) current documents which are usually short-lived, e.g. invoices or supply indents. In either

11

case much unnecessary work can be avoided if the records are carefully designed at their inception.

In a large organization the figures initially compiled from the basic records may have to be submitted to a central department for incorporation with others. For this purpose a form of return is normally provided. This also should be designed with care to ensure that it is easy to complete correctly and to interpret when received. The form taken by these records and returns will depend on such factors as their ultimate use, the unit period covered by each return, the amount of detail and the degree of accuracy required, and the permissible delay.

Permanent records, particularly if they are also the source of statistical data, should normally be on cards or in loose-leaf form, so that it is easy to insert new ones or to sort them into any classification which may be required. If it is not desirable to sort the cards (because they are subject to constant reference) or if the amount of information to be recorded necessitates a large card, the "visible index" type of card is to be recommended. These cards are filed, usually in trays contained in cabinets, so that one line of each card is visible when a tray is pulled out. This is known as the visible strip, and essential information can be entered either in full or in code, or coloured signals can be attached to indicate certain particulars of the customer or employee to whom the card relates. Thus, any customer whose turnover with the company exceeds £1,000 per annum might be indicated by a red signal; at any time it is easy to count up the number of customers in this category by pulling out each tray in turn and simply counting the number of red signals. If cards are frequently removed from the trays, separate visible strips may be used which remain in position when the cards are extracted.

When current documents form the source of routine statistics it is generally necessary to transfer the required information to a summary sheet. It is easy to record far too much information for this purpose. For example, a Sales Manager might require up-to-date daily statistics of the total volume of each product ordered. This would be obtained by recording each order as it was received, and it would be sufficient to enter the quantity ordered in the column appropriate to the product. Details such as the customer's name, the order number and the packages involved might be entirely irrelevant and should, therefore, be rigidly excluded from the record.

A record card or return should be so designed that it is straightforward to complete and the descriptions and instructions used are unambiguous. If, for lack of space, it is not possible to make it fully self-explanatory, then precise instructions must be given to those responsible for completing it, and a check should be made soon after a new record or return is instituted to make sure that it is being completed correctly. Thus, when a statement of absence is required, it should be made clear whether this is to include very short periods of absence, such as lateness, or to be confined to periods of not less than a whole day or shift; similarly, the expression "days lost" without further explanation might mean only working days, or it might include week-ends and holidays. When calculations are required, e.g. percentages, the totals concerned should be fully defined, and it is useful also to indicate the degree of accuracy which is required, otherwise some returns will show the percentages carried to numbers of decimal places which are meaningless.

Routine statistics usually relate to the past; their significance lies in the extent to which they affect actions or decisions in the future. It is therefore important, in designing a system of routine statistics, to be clear about their ultimate purpose, for it may be possible to select the significant information and present that alone on the final statement. For example, a manager may require sales statistics for the purpose of detecting changes in the pattern of sales, and he is less interested in the figures which accord with expectations. In this case actual sales can be compared with forecast sales, and only those which differ by more than a specified amount need be presented to the manager. He knows then that every figure in the statement is of some importance, and he is saved the task of ploughing through a mass of figures on which no action is required.

Coding

It is often convenient to code the information contained on a card or form, using letters or numbers to represent the various classifications. For example, staff functions may be coded as follows—

COMMERCIAL—

Administration	00
Distribution	01
Purchasing and Supply	02
Sales	03
Transport	04

PRODUCTION—

Works Management	10
Maintenance	11
Planning and Progress	12
Supervision	13
Instruments	14

and so on. Instead of writing "Distribution" the clerks simply write "01" in the appropriate space. This saves time and space and makes it easier to sort the cards and tabulate the information. The advantage of a double-digit code, when, as in the example quoted, the classification consists of a number of main headings and a more detailed subdivision, is that a tabulation may be either general or detailed according to its purpose; for the general one it is sufficient to sort on the first digit, but for the detailed one it is necessary to sort on both digits. Coding is usually necessary when punched cards are used (see below).

Punched Cards

Cards into which holes are punched to denote the required information by their position are of two kinds: those used in connexion with computers or with electrical or mechanical sorting and tabulating machines, and those which are sorted by hand with the assistance of special devices. The equipment used and the method of operation for the first type are dealt with in Appendix 1. With the other type of punched card, holes are provided either round the edge or in part of the body of the card, and each hole represents a code number; when the information (which may also be written on the card) has been coded, the appropriate holes are elongated by a special instrument and the card is filed. When it is desired to select the cards relating to a particular code number, a needle is passed through the appropriate hole, through the pack of cards, and those which have been punched in this position either fall out or protrude according to the particular type used.

Punched-card machinery is quite expensive, and a substantial volume of work is required to justify an installation. It is probably most economical when the punched cards recording the basic data can be used for a variety of purposes, e.g. preparing invoices on the tabulating machine as well as providing sales statistics. It is very valuable also where a number of classifications are required from the same basic data. For example, from a set of punched cards each representing one order it would be possible to produce all kinds of tabulations for various purposes, e.g. quantity and value of each product purchased by each customer, customers purchasing each product, numbers of each type of package used, quantity of each product consigned to each town or county, and so on. Care should be taken, however, to ensure that the mere possession of a machine capable of producing all this information does not in itself lead to a demand for unnecessary tabulations.

The hand-sorted punched cards are useful where it is occasionally necessary to extract statistical information but the job does not justify expensive equipment. For example, it might be convenient in a moderately large organization to record particulars of each accident on a punched card and periodically to examine all accidents due to failure to observe safety regulations, or those which resulted in some particular injury such as a broken leg.

Sales Statistics

Sales statistics merit prior consideration, for it is the demand for a company's products that sets in motion the whole complex machinery of production. Sales statistics, which consist essentially of the quantities of products supplied, may be compiled either from the orders as they are received or from the internal records subsequently prepared, e.g. copy invoices. Comparison with finished stocks and the production programme will show whether all is well or action is necessary to stimulate demand, e.g. by intensified advertising, or even to reduce it by a quota system, and whether any change in the production programme is needed.

Sales statistics in detail may show the sales to each customer (and indicate the attention to be given by representative or the credit terms to be allowed), the sizes and packages required (to assist planning in the Packaging Department), and the volume of goods dispatched to various parts of the country (to determine the most economical method of dispatch to various places). They may also provide information on the ultimate use of the company's products by various industries and contribute to Government statistics on this subject. Sales statistics sometimes include details of inquiries received which have not resulted in orders, and "competition intelligence," i.e. information on the proportion of a customer's trade held by a competitor.

Sales statistics are often required in two stages—"red-hot" daily figures (which need not be exact) for immediate sales action and production planning, and "historical" exact figures for accounting purposes and long-term planning. The unit period may be a day for the first purpose and three months for the second; the previous day's figures may be required first thing each morning, but it may be immaterial if the quarterly return is not available until a month after the end of the period.

It is very easy to produce too much detail in sales statistics, and this can be very costly. Hence, it is essential that the exact use of the figures should be decided at the outset so that any irrelevant detail can be eliminated. It is also easy to produce exact statistics too often. A period such as a week, or even a month, may well be too short for any important trend in demand to reveal itself, and returns based on such a period may lead to waste of time in trying to find the reasons for fluctuations which are simply normal.

Sales forecasts are usually prepared periodically based on previous experience, market trends and surveys, and general information obtained by representatives from customers. The forecast may be quite simple in terms of total expected sales, or it may be detailed by localities or even down to individual large customers. The value of a forecast is that it provides a standard by which actual achievements may be judged, and it facilitates the detection of any failing or shortcoming in any part of the selling force.

Selling expenses may be analysed to ensure that the selling force is economically deployed and that advertising expense is directed into the most fruitful channels. There is some difficulty in allocating selling expenses, particularly when a representative is dealing with several products, and when an attempt is being made to build up a new market; some approximation may be necessary and the analysis should be treated with caution.

Production Statistics

Production statistics relate to the quantity and quality of output achieved, which is usually compared with the production plan or "budget," and to such matters as machine running time, individual workers' or group performances, numbers and duration of breakdowns, and the quantity and value of spoiled material. This information is generally obtainable from workshop orders which detail the work to be done, or from records used to allocate workers' time over their various jobs.

The purpose of production statistics is to enable the Production Manager to ensure that he is fulfilling the demands of the Sales staff and operating economically. This again illustrates the essential function of statistics—control. The foreman "knows it all"—he knows his men, he knows the work going through, he knows when a machine breaks down, and he does not need routine statistics to tell him these things. The Manager, however, cannot know it all and must depend on the routine statistics which are supplied to him.

Supply Statistics

Supply statistics will show primarily the quantity and value of the stocks of materials, components and equipment held. By comparing changes in stock levels the Supply Manager can see whether his buying policy is economical, and he can act to prevent either over-stocking or under-stocking. He is concerned to avoid tying up too much money in stocks, but at the same time to avoid running out of some vital component and thereby possibly causing a plant to stop production. By examining the fluctuations in demand for various items and seeing how long suppliers take to deliver, he can fix stock levels at which an order should be placed and the minimum level at which special action should be taken to replenish stocks. The basic record is usually a stock card for each item, on which the receipts and issues are entered; this card often provides for periodical calculation of the value of stocks held which can be reconciled with the accounts.

More detailed analyses can assist the Supply Manager further. The comparison of transport charges will help to determine the most economic mode of delivery for the principal items of stock. Evaluation of stocks of certain small items may show that the administrative cost of maintaining detailed records for these is unjustified and may lead to a reduction of clerical work. Statistics of orders typed will help to show whether the right size of staff is employed and whether there are seasonal fluctuations in the work. In these ways the routine statistics can give the Supply Manager the control which he needs.

Personnel Statistics

Personnel statistics yield information about the staff and labour force employed, wages and salaries, hours worked, absence and turnover. Detailed records of the behaviour of a large group of employees can be a very useful source of information for guiding personnel policy, but in this field particularly so many possible classifications can be made that there is a great need for careful selection if the final statements are not to be so complicated that they conceal the salient features. The basic employee record usually includes the essential details of the employee himself, i.e. name, address, sex, age and possibly whether married, and number of dependants, particulars of his education, training and previous employment, and information about his present employment with the company, i.e. date engaged, occupation, results of any tests, and so on. Changes of occupation and status and other matters of major importance will be recorded on this card as time goes on. Particulars of his timekeeping and absence record may be included or these may be more conveniently compiled on a separate record card.

In personnel statistics one rule that must be observed is to keep figures for men separate from those for women. Apart from this, the classification is largely a matter of choice, but it should be possible to analyse employees at any time according to their grade or occupation, factory or department, and it is desirable to be able to sort them into age groups and possibly into length of service groups. Division into married and single is more important for women than for men, but it may be important for men if there is a pension scheme.

Two problems of great concern to most large organizations are absence and labour turnover (i.e. numbers leaving the employer's service). It is not easy for an individual firm to influence either of these factors, but they can be so wasteful that it is well worth trying to analyse their causes. Sickness should be analysed by sex, category of workers, etc., and if possible by age group and diagnosis. A thorough analysis involves coding and, if possible, the use of punched cards or a computer. For diagnoses there is a three-figure International Statistical Classification of Diseases and Injuries compiled by the World Health Organization.

Labour turnover can be divided into voluntary (i.e. those who leave of their own accord) and non-voluntary (i.e. resulting from retirements, deaths and dismissals). It is notable that voluntary turnover is generally much heavier in the first few months of employment than later on, and an apparently high turnover rate may in fact be due to the "fringe" of workers (perhaps 10 per cent) who are constantly changing their jobs; this can be an expensive business for the employer because of the administrative and training costs involved in the engagement of a worker. It is difficult to ascertain the real reasons for workers leaving, but it may often be done by skilful interviewers if they are given sufficient time. Turnover can, therefore, be analysed by sex, married or single (at least for women), length of service, age, grade or type of work and, very broadly, by causes of leaving.

On thing that must be stressed here is the importance of being able to compare any figures of sickness or turnover with the corresponding numbers exposed to risk. For example, if the number of workers leaving with less than three months' service falls from 500 to 300 in successive quarters, it may only mean that fewer men were engaged in the one quarter than in the preceding one. No valid deduction can be drawn from the numbers leaving unless the average number of employees in the category chosen can be calculated for each period. It is this sort of point that must be borne in mind when returns are designed.

4

Censuses and Surveys

THE previous chapter dealt with routine statistics collected for the continuous information and guidance of management, often as a by-product of administration. The present chapter is concerned with inquiries designed to answer specific questions or to provide certain information once and for all.

A census is generally a complete count of a population (or shops, farms, etc.), whereas a survey is usually conducted on a sample basis and aimed at obtaining more general or more detailed information. The distinction, however, is by no means clear-cut, as is shown by the "Sample Census of Population" in 1966 (see Chapter 18). Moreover, a census is generally much more than a mere enumeration, and may be used to obtain information about such matters as housing and car ownership.

Full-Scale or Sample Inquiry?

With a fairly small and preferably intelligent population, there is little difficulty. When the Institute of Statisticians conducts a Salary Survey, it sends a questionnaire to all its members with every confidence that the great majority of them will complete it correctly and return it promptly. When a market research organization wants to conduct a nation-wide inquiry, embracing all types of people, that is another matter. Such an inquiry is only practicable on a sample basis.

The principle underlying sampling is that a set of objects taken at random from a larger group tends to reproduce the characteristics of that larger group. This is called the Law of Statistical Regularity. There are exceptions to this rule, and a certain amount of judgment must be exercised, especially when there are a few abnormally large items in the larger group. With erratic data, the accuracy of sampling can often be tested by comparing several samples. On the whole, the larger the sample the more closely will it tend to resemble the population from which it is taken. Too small a sample would not give reliable results. It will be shown later in Chapter 13 how the precision of estimates obtained from a sample can often be estimated from the sample itself.

The Law of Statistical Regularity may be exemplified by assurance statistics. Actuaries base their life tables and calculations of premiums to be charged on masses of data obtained from past experience, extending over many years. Experience shows that mortality rates do not vary greatly from one year to another, although they certainly have fallen over a long period and are still falling. Furthermore, the experience of all companies in any one year is practically the same, so that although the records of any one company are only a sample, they are clearly representative of the population as a whole, with the proviso that a small section of the population is uninsurable.

The advantages of a sample are that it costs much less and takes less time than a full-scale inquiry, and that the results are known more quickly; also that it is possible to ask more questions and obtain more information than would be possible in a full-scale census. If the inquiry is made periodically, sampling enables it to be made more frequently. On the other hand, a sample nearly always means some loss of accuracy, but even here there are exceptions; a sample survey carried out with due care by trained people may give better results than a full-scale census taken indifferently.

The decision whether to take a sample therefore depends mainly on the cost of the inquiry in time and money, the delay in obtaining the results if a full-scale census is chosen and the degree of accuracy required. If, for example, the information is already available on punched cards which can be quickly sorted and tabulated, a full census is generally advisable. This might apply, for example, to an investigation into employee turnover in a large firm.

The Problem of Non-Response

When it is necessary to ask people for information, and particularly when this is done by a postal inquiry, it often happens that a large proportion of those approached do not reply. It is then difficult to judge whether those who do are representative of the whole population. It may be that most of the replies come from people who have an axe to grind or feel strongly on a particular issue. If so, the results of the survey will clearly be biased.

This problem of non-response, or partial response, does not arise with observations on animals, plants or inanimate substances, but it is a very troublesome problem in market or opinion research.

It can occur with full-scale as well as sample surveys. Thus, if all members of a university are sent a questionnaire about their sex life, or all employees of a firm are asked for their views on a matter of company policy, and only 20 per cent reply, that 20 per cent may contain a high proportion of cranks or misfits.

In order to encourage a good response, the questionnaire should be made as simple as possible, and accompanied by a good covering letter explaining clearly why the survey is being conducted and why it is important for the recipient to complete the questionnaire, and stressing that replies are strictly confidential. A pre-paid business envelope should be enclosed and a reminder card can be sent if necessary. The replies received after the reminder has been sent should be compared with those received earlier. If they differ, it can be assumed that those who have still not replied resemble the second batch rather than the first.

Mail Inquiry or Interview?

A postal inquiry is an easy and inexpensive way of obtaining information, and for routine business and administrative inquiries it is the standard method, but in market research, public opinion surveys and similar inquiries, where a high response rate is essential, personal interviews are usually best. They are generally conducted by trained "field staff" armed with a standard list of questions. Few people will refuse to co-operate when personally approached, so this method ensures a fairly complete response. Moreover, the interviewer can explain any questions that are not clear to the person interviewed, but he (or she) must be careful not to suggest the answers or to bias the results in any way. This is not so simple as it looks. "Interviewer bias" can easily arise in political or similar inquiries through the interviewer misunderstanding a reply or marking a wrong code number on his schedule. Replies can also be biased through forgetfulness on the part of the people interviewed, by the desire to make a good impression on the interviewer (the "halo effect"), or by the fear that a truthful answer may result in something to their disadvantage. These are some of the principal "response errors." Thus, while the interview obviates most of the bias due to non-response found in a postal inquiry, it introduces other possible sources of error.

The interviewer must, of course, find his people before he can interview them. This is time-consuming, and some will be difficult to contact. Many will be persistently out. Some may be night workers, or away on business or on holiday. Others may be too ill or too deaf to be interviewed. Interviewing therefore has its own drawbacks, besides being expensive in salaries and travelling costs.

The Questionnaire

The importance of a good questionnaire can hardly be exaggerated. Even in an interview, where the interviewer reads the questions and himself records the answers, it is essential that the questions should be clear, framed in simple language, free from ambiguity, in logical order, and such that the respondents can and will answer them honestly and intelligently. Leading questions ("Do you agree that . . .") and emotive words ("big monopolies") should be avoided. Bias can be caused just as much by loaded questions as by non-response.

It is a common practice to run a *pilot survey* before the main inquiry, primarily to test the questions. For this purpose a relatively small sample is chosen, which need not be a representative cross-section of the population, since at this stage the actual answers are unimportant. A certain amount of information may emerge, such as an estimate of the likely non-response or differences between different groups of people, but the main object is to see whether the questions and instructions are foolproof, and the whole questionnaire well designed.

Selecting the Sample

In order to select the sample, it is usually necessary to have the whole population listed in some way. An employer will have a file, or at least a card, for each employee, arranged alphabetically or in some other logical order. A firm may have a mailing list of actual or potential customers. Such a list, or set of lists, maps or similar records, is called a *sampling frame*. The two most commonly used sampling frames in Great Britain are the Electoral Register (for persons) and the Rating Lists (for dwellings).

For a sample to be truly representative, it is essential that the sampling frame should be complete, up-to-date and adequate for the purpose. A public opinion survey should not be taken from one particular section of the population, e.g. those whose names are in the telephone directory, or readers of a particular newspaper, unless it is made perfectly clear that the survey relates only to that section.

There are various ways of selecting the sample, such as

(*a*) Random sampling
(*b*) Systematic sampling
(*c*) Stratified sampling
(*d*) Cluster sampling
(*e*) Quota sampling

There are a number of variations and combinations of these methods, but these are the principal ones, and they will be described briefly in turn.

(a) Random Sampling

Random sampling simply means choosing a sample in such a way that at each step all members of the population not already selected stand the same chance of being chosen. Thus a hand of cards dealt from a properly shuffled pack is a random sample. Winning tickets in a lottery are chosen at random by taking numbers out of a revolving drum. By a simple extension of this idea, a random sample can be taken from a group or population by numbering all items of the group and picking numbers by means of a table of "random numbers," i.e. numbers of (say) two or four digits compiled by some random process.

Note that "random" does not mean "haphazard," just thinking of a number. The method must be completely objective. If you doubt this try asking a large class of students to write down a digit at random—any digit from 1 to 9. Probably about 30 per cent of them will put down "7". There is a bias against 1 and 9, and possibly 5, because they do not "look random," but no one really knows why 7 is so popular. Provided the sampling frame is sound, a random sample will be completely unbiased. This cannot be said of any other method.

(b) Systematic Sampling

This means taking the sample in some systematic manner, generally by taking items at regular intervals. Thus, in an agricultural experiment, a plot of land could be divided into strips of equal width, and every tenth strip selected. The same method can be used in sampling from a list, provided there is no systematic arrangement in the list, or that the order of the list has no bearing on the subject of the survey. A 10 per cent sample survey of sickness absence has been conducted in a large organization, taking every worker whose Works Number ended in a specified digit. Another possibility would be to take every person whose birthday occurred on a particular date in the month. Such samples are sometimes called *quasi-random*, because they are as good as random for the purpose in hand.

(c) Stratified Sampling

This method makes use of prior knowledge of the population, and is the most commonly used type of sample design. The sampling frame is first divided into natural sections or *strata* (plural of *stratum* so, whatever you do, avoid talking about "stratas"). This will probably be done already, if the frame is the Electoral Register or some other records held by Local Authorities scattered over the country. The strata may be regions, towns, streets, sexes, age-groups, social classes, occupations, and so on, depending on the nature of the inquiry. Within each region, a random sample of towns, rural districts, etc. may be chosen. In fact, sampling may be done at several stages (multi-stage sampling). Finally, within each stratum or sub-stratum selected, a random sample of the required size is chosen.

There are many advantages in stratification. It ensures that each section of the population is adequately represented and that there is sufficient information about that section. It is less expensive than pure random sampling, because field staff can concentrate on a limited number of areas instead of having to visit one man in Skye or in the middle of Dartmoor. It is necessary, however, that the proportions of the population in the various strata should be known beforehand, at least fairly accurately.

(d) Cluster Sampling

This is a variation of stratified sampling where, instead of taking a random sample within each sub-stratum selected, the whole of the sub-stratum is taken. It might, for example, be every child in a class or every member of a household. It is, in fact, a multi-stage sample with 100 per cent sampling (if that is not a contradiction in terms) at the last stage.

(e) Quota Sampling

This might be described as non-random stratified sampling. The total sample within each selected area is broken down according to sex, age group and social class, and the interviewer is given complete discretion as to how he chooses the individuals. This procedure has obvious dangers, as shown by the following incident. A survey was being conducted into betting habits. An interviewer arrived in London early one morning and decided to take some interviews before going to the office. What better than a queue of people waiting for a train, with nothing to do? Looking at the results afterwards he found they were hopelessly biased, and on making inquiries, discovered that the train was bound for Newmarket Races.

Quota sampling is cheap and easy, requires no sampling frame and no call-backs, and is very useful in certain types of inquiry, but it is biased towards people working out of doors and people with a good education. The judgment of social class is obviously very subjective. The main drawback, however, is that whereas it is possible to put estimates of reliability (see Chapter 13) on the results of a simple random or stratified random sample, this is not possible with quota sampling.

Suggestions for further reading

MOSER, C. A. and KALTON, G., *Survey Methods in Social Investigation* (Heinemann)
YATES, F., *Sampling Methods for Censuses and Surveys* (Griffin)
KISH, L., *Survey Sampling* (Wiley)
ILERSIC, A. R., *Statistics* (HFL (Publishers) Ltd)

Exercises

4.1 A large company wishes to gather information about the newspaper reading habits of the general population, in order to find any correlation between reading habits and other characteristics such as age, sex, education, etc. Design a questionnaire, to be used postally, to give such information as you think appropriate. (IoS)

4.2 Write a brief explanation of the following terms used in market research

    (*a*) sampling frame;
    (*b*) interviewer bias;
    (*c*) non-response rate;
    (*d*) non-sampling errors. (ICWA)

4.3 Statistical information may be obtained by the use of questionnaires. What steps should be taken to ensure that

    (*a*) the questionnaire is adequately designed, and
    (*b*) that the method of collection (by post or direct interview) does not influence the results? (ACCA)

4.4 Before any large survey a pilot survey is usually carried out.

    (*a*) Why is a pilot survey needed?
    (*b*) What can it reveal?
    (*c*) Who should carry it out?

Answer in note form. (IoS)

4.5 A manufacturing company sent several interviewers out on to the streets with instructions to each to ask 50 people what they thought about the company's products. If this was all the instruction given to the interviewers, what sort of errors and bias would appear in the answers? Answer in note form, giving a brief example to illustrate each point. (IoS)

4.6 Why do statisticians often use sampling techniques to obtain information? What are the purposes and procedures involved in random, systematic and stratified sampling? (ACCA)

4.7 A manufacturer of gas appliances wishes to carry out a survey of households in a given town in order to assess the success, or otherwise, of their recent conversion to natural gas.

    You have been asked to advise on the following points

    (*a*) How big a sample should he pick?
    (*b*) What should he use as a sampling frame?

(c) Should he use postal questionnaires or carry out interviews?

(d) Should he use open-ended or precoded questions?

Answer each of the questions briefly, stating in each case the considerations that have led to your decision.   (ICWA)

4.8 A market research agency was asked to interview as many women as possible among 1,200 married women selected at random from a voting list for a district in a large town. The interviews were to be completed in a week and each woman was to be asked whether she had heard of a cake-mix called "Fairymix." If the reply was "Yes," the respondent was to be asked if she had a packet in her larder.

The results were as follows

Number interviewed	550
Number who knew of Fairymix	100
Number who had a packet	23

Comment on the usefulness of this survey to the manufacturers of Fairymix.   (IoS)

5

Statistical Tables

WHEN the statistician has collected his data in the form of cards, schedules or other documents, he may be faced with an amorphous mass of figures which must somehow be crystallized into a reasonably compact statement. Both the final product and the intermediate stages will probably consist largely, if not entirely, of statistical tables, since they provide a convenient way of summarizing the data in an orderly manner and of presenting the results concisely and intelligibly.

Structure of Tables

A table is a rectangular arrangement of figures, with headings describing the various rows and columns. The first column, containing the headings for the rows, is called the *stub*, and the headings of the other columns form the *captions*, while the figures constitute the *body* of the table.

Table 5.1 Mark-sheet of examination candidates and their marks
(An example of two-dimensional tabulation)

No	Name	English	French	Mathe-matics	History	Total
1	Abbott, T. R.	54	62	37	55	208
2	Ackroyd, B. J.	71	58	70	64	263
..	. . . . .	.	.	.	.	.
113	Young, C.	67	48	82	63	260
	Total	6,392	6,205	5,977	5,649	24,223

27

The simplest form of table is a single column of figures with the stub, like a list of examination candidates and their marks in a single subject. The term "table" however, generally conveys the idea of a set of figures arrayed in at least two dimensions, such as a mark-sheet giving the candidates' marks for a number of subjects, with totals, as in Table 5.1.

Notice that there are two marginal sets of totals, those for each candidate and those for each subject. The former are, of course, essential. The latter might be thought unnecessary, but they provide a useful check on the accuracy of the last column because the sum of the subject totals and that of the candidates' totals should agree in the same grand total. It is always desirable to make a table self-checking whenever possible.

Tabulation in three or more dimensions is not possible in the literal sense, but the same effect may be obtained by subdividing rows or columns or both. Table 5.2 is an example of tabulation in three dimensions, each of the main column headings being sub-divided into Male, Female, and Total.

Table 5.2   Summary table showing numbers employed, total earnings and hours worked in Brown and Green Ltd in the week ended 6 March 1971

(An example of three-dimensional tabulation)

Factory	Nos employed			Total earnings			Man-hours worked		
	M	F	T	M	F	T	M	F	T
				£	£	£			
A	150	46	196	3,330	742	4,072	6,681	1,871	8,552
B	228	143	371	5,827	1,708	7,535	9,802	6,114	15,916
C	570	269	839	15,427	4,195	19,622	27,719	12,325	40,044
D	251	73	324	6,074	967	7,041	11,348	3,126	14,474
Total	1,199	531	1,730	30,658	7,612	38,270	55,550	23,436	78,986

In Table 5.3 the process has been carried a stage further. Again the main column headings are subdivided, this time into years, but the row headings for commodities are also subdivided, not only into more detailed commodities but in some cases into country of destination. Both kinds of sub-headings are shown slightly indented to the right—too slightly to make it clear which are sub-headings of what. The compilers have tried to compress the maximum of information into the page, thus suffering some loss of clarity. It is only fair to say that the Trade Accounts are no longer published in this form.

Table 5.3 Extract from UK trade accounts for January 1953
(An example of four-dimensional tabulation, now happily obsolete)
II—Exports (produce and manufactures of the United Kingdom)

Class III		Quantities			Value		
Articles wholly or mainly manufactured		Month ended 31 January			Month ended 31 January		
		1951	1952	1953	1951	1952	1953
GROUP F.—ELECTRICAL GOODS AND APPARATUS—							
Electric cables, wires, strips and strands, insulated—							
Telegraph and telephone cables and wires—					£	£	£
Submarine	Tons	263	436	75	35,921	60,234	12,171
Other—							
To Union of South Africa	,,	228	154	356	57,748	53,313	129,571
To India	,,	40	67	59	11,595	25,393	39,210
To Malaya	,,	204	25	68	42,670	10,467	18,259
To Australia	,,	1,118	130	82	248,833	49,159	39,045
To New Zealand	,,	315	470	247	77,728	129,574	104,162
To Other Commonwealth Countries	,,	352	304	382	76,858	98,831	114,687
To Irish Republic	,,	218	36	27	49,872	11,777	21,386
To Argentine Republic	,,	98	19	—	32,010	9,767	—
To Other Foreign Countries	,,	122	422	480	43,944	153,215	178,279
Total		2,695	1,627	1,701	641,258	541,496	644,599
Other descriptions—							
Paper insulated—							
To British West Africa	,,	127	67	31	26,259	15,015	9,383
To Union of South Africa	,,	429	105	244	77,339	27,435	67,763
To India	,,	417	622	1,469	75,647	136,074	382,802
To Malaya	,,	127	122	392	22,876	30,606	104,725
To Hong Kong	,,	18	63	71	3,401	16,253	16,878
To Australia	,,	751	745	936	137,731	163,399	267,035
To New Zealand	,,	208	58	79	38,172	16,012	22,930
To Other Commonwealth Countries and the Irish Republic	,,	568	492	1,252	100,613	124,032	307,030
To Norway	,,	34	32	—	6,793	7,620	—
To Portugal	,,	30	18	11	5,329	3,748	2,715
To Turkey	,,	—	147	181	—	32,733	43,374
To Egypt	,,	4	122	17	730	30,359	5,459
To Other Foreign Countries	,,	272	571	327	48,546	141,270	99,826
Total	,,	2,985	3,164	5,010	543.436	744,556	1,329,920
Rubber insulated—							
To Union of South Africa	,,	201	163	128	61,440	71,931	55,560
To British East Africa	,,	70	97	26	26,856	39,127	13,358
To India	,,	243	151	190	57,972	61,644	76,020
To Pakistan	,,	95	91	9	30,359	53,306	3,158
To Malaya	,,	57	150	98	15,863	71,692	35,870
To Hong Kong	,,	150	60	61	41,531	26,420	25,235
To Australia	,,	27	59	31	14,662	24,979	11,998
To New Zealand	,,	199	233	67	54,161	116,034	28,759
To Other Commonwealth Countries and the Irish Republic	,,	205	277	353	65,306	150,751	158,529
To Foreign Countries	,,	238	300	263	90,268	145,583	129,689
Total	,,	1,485	1,581	1,226	458,418	761,467	538,176

Too complicated classification and subdivision of headings make a table difficult to read, and it is generally better to break it up into a number of separate tables or to condense the information.

Stages in Tabulation

The amount of tabulation necessary varies considerably according to the nature of the inquiry and of the data available. A typical case, involving the analysis of a number of schedules bearing information on several points, might contain the following stages—

1. Extracting the relevant items of information from the schedules and entering them on "working sheets" designed to facilitate calculations.

2. Performing the necessary calculations with the aid of these working sheets.

3. Transferring the results of these calculations—totals, etc.—to one or more summary sheets and subjecting them to further manipulation.

4. Compiling one or more summary tables embodying the results in all necessary detail, but omitting original and intermediate figures which have served their purpose. These tables will probably include derivative data such as averages and percentages. For example, a summary table derived from Table 5.2 might give average weekly earnings and hours worked in each factory and in the firm as a whole, together with percentage increases in average earnings since a given date.

Working Sheets

Working sheets are generally kept as a permanent record of the calculations performed on the data, both as a guide for future occasions and for reference in case figures are queried or further detail is required. They should therefore be drawn up carefully on good paper that will not tear easily. It is often worth while spending considerable time on designing a working sheet so that the calculations can be done as efficiently as possible. If many similar working sheets are required they should be stencilled or otherwise reproduced with all possible detail such as headings, footnotes, units, £ signs, and cross-references between items, saving the compiler both time and thought.

Plenty of room should be allowed for amendments and corrections. Sheets should be given proper titles, and columns their proper headings; continuation sheets should be numbered consecutively. Units should be clearly shown, and if necessary the source of the data should be stated. As mentioned earlier, whenever possible the calculations should be self-checking. A control column can be introduced if necessary to provide a check.

When a large number of items have to be counted, a useful device is to make a vertical stroke for each of the first four items and a stroke across them for the fifth, forming a gate. Thus, 37 items would be recorded as follows—

卌 卌 卌 卌
卌 卌 卌 ||

This "gate system" makes counting much easier and more accurate than a mere succession of strokes.

Hints on the Construction of Tables

1. Having thought out what the table is to show, give it a clear and comprehensive title. Be particularly careful about dates and geographical regions. Some official figures, for example, refer to the United Kingdom, others to Great Britain, and others to England and Wales. If the table covers the years, say, 1960–1970, this can be shown in the title, e.g.—

UK PRODUCTION OF PIG IRON, 1960–1970

2. Consider the column and row headings carefully, and make sure they express exactly what is intended. Show totals and subtotals clearly, if necessary by heavy type for the relevant headings and figures or by light and heavy rulings, and show sub-headings in the stub by judicious indenting. It is sometimes desirable to show totals at the top or in the first column of the table, particularly if only some of the component items are shown and they do not add up to the total.

3. State the units clearly, at the head of each column, or on the right-hand side of the stub (see Table 5.3) or, if all figures in the table are in the same units, above the whole table. If figures are given to the nearest hundred, thousand, or million, the unit can be given as thousand tons, £ million, etc. Much space and labour can be saved by working to three- or four-figure accuracy.

4. Make sure that comparisons shown in the table are valid. For example, output in periods of sometimes four, sometimes five weeks should be shown in weekly averages, and in presenting annual and quarterly figures side by side, the quarterly figures can often be given as equivalent annual rates. Sickness rates for different groups of workers should not be shown for comparison without comment if the age-compositions of the groups are very different.

5. Give the source of your information at the foot of the table where necessary. Nothing is more exasperating than to be unable to trace the source of a table. If it is a regular publication, give the date of it; figures may be revised in a later edition.

6. Any definitions or explanations that cannot be conveniently shown in the table should be given in footnotes or other accompanying notes, with numbers, asterisks or other symbols against the items or headings to which the notes refer. Any break in continuity through changes in classifications, etc., must be clearly shown.

7. Avoid complicated tables if you can, but if not, it may be helpful to number the rows or columns. The relations between them can then be shown simply, as in the following example—

GROSS OUTPUT	RAW MATERIALS	NET OUTPUT
(1)	(2)	(3)=(1)−(2)

With a very wide table, repeat the row headings or numbers on the right-hand side.

8. With a long table, e.g. one involving a long list of countries, leave a space after every fifth item. With monthly data it may be preferable to give blocks of three or six months, thus—

1970: January
February
March

April
May
June

9. The modern tendency is to reduce vertical lines in typed or printed tables to a minimum, e.g. to put them between main headings but not between sub-headings in a table like Table 5.2. If figures are to be written by hand it is best to draw vertical lines to keep the columns straight.

Some of the above hints apply equally to working sheets. The list is by no means exhaustive, but the student who wishes to learn how to draft tables cannot do better than study the *Monthly Digest of Statistics* and other official publications.

Exercises

5.1 In his annual statement, the Chairman of a group of three companies, *A*, *B*, *C*, gave the following analysis of the profit (in £'s sterling) from trading in various parts of the world. "For Company *A* the total profit was £130,000; of this sum, £100,000 came from the United Kingdom and £10,000 from trade with the Commonwealth, whereas profit in Europe amounted to £3,000 from the European Free Trade Area (EFTA), together with £15,000 from the European Economic Community (EEC); profit from the USA was only £2,000. As for

Company *B*, this company made £67,000 in the United Kingdom but had no trade with the USA; profit from EFTA and EEC countries was £1,500 and £5,000 respectively which, together with £2,500 from the Commonwealth, made a total of £76,000. Finally, Company *C* made the lowest total profit £52,800; of this, £40,000 was made in the United Kingdom, the Commonwealth profit being £5,700 compared with £2,100 from EEC and 5000 from EFTA."

Tabulate these data, adding any secondary statistics which may be helpful.   (IoS)

5.2

Road and rail goods transport in Great Britain

	1958		1964	
	Road	*Rail*	*Road*	*Rail*
Weight of goods carried (mn. tons)	1,061	243	1,410	240
Ton miles (thousand million)	25·2	18·4	39·0	16·1
Average cost of transport:				
per ton (£)	0·99	1·07	1·43	0·97
per ton mile (£)	0·042	0·0141	0·052	0·0145
Average length of haul (miles)				

Calculate the figures needed to complete the above table. Discuss what the completed table tells you about the relative nature of road and rail goods traffic in Great Britain and the changes that occurred between 1958 and 1964.   (IoT)

5.3 In a market survey, housewives were interviewed to determine their preferences for, and use of, detergents. 100 housewives over 40 years of age were interviewed, and 150 under 40; 80 of the first group and 130 of the second group said that they used some kind of detergent. In answer to a second question—"Do you use detergents for

   (*a*) Clothes washing
   (*b*) Household purposes (e.g. washing floors, etc.)
   (*c*) Other uses (e.g. car washing)?",

50 answered "Yes" to part (*a*) from the over-40 group and 70 of the under-40's said they used detergents for clothes washing. In answer to parts (*b*) and (*c*), 70 and 5 respectively answered "Yes" in the over-40 age group. The corresponding figures for the under-40's were 100 and 10.

Tabulate the data, adding any secondary statistics which may be helpful in illustrating the results of the inquiry.   (IoS)

5.4 Between the end of 1950 and the end of 1964 the number of United Kingdom vessels of 500 gross tons and over decreased by 619 whereas the gross tonnage increased by 3,229,000 tons. Of the total number of vessels at the end of 1964, 213 were passenger vessels and 535 tankers. The remainder were dry cargo vessels, of which there were 452 fewer than at the end of 1950. The 563 tankers in service at the end of 1950 had a gross tonnage of 3,947,000 tons, the corresponding gross tonnages for passenger and dry cargo vessels being 2,937,000 tons and 10,315,000 tons respectively. The gross tonnage of passenger vessels at the end of 1964 was 2,244,000 tons, compared with a gross tonnage of 10,380,000 tons for the 1,725 dry cargo vessels in service.

Draw up a table which shows, at the two dates, the number, gross tonnage and average tonnage for each type of vessel and for all merchant vessels. Write brief comments on the main changes between the two dates as revealed by your table.   (IoT)

5.5 "The industry's actual capital expenditure on fixed assets of £294·9 million in 1960–1961 came very close to the estimate of £291·8 million approved by the Minister in August 1960. The Generating Board spent £196·2 million and the Area Boards £98·7 million, against approved estimates of £196·5 million and £95·3 million, respectively. The Generating Board's expenditure included £53·5 million for nuclear power stations, expenditure on nuclear fuel representing £3·2 million which was £3·3 million less than the approved estimate of £6·5 million."

Tabulate these data, deriving any secondary statistics which may be helpful.   (IoS)

5.6 In 1958, 1,090 million passenger journeys were made on British Railways, yielding £137·6 million in receipts. Of the journeys 351 million were at full fares, 426 million at reduced fares and an estimated 313 million by season ticket holders. By 1961, journeys at full fares had fallen by 78 million, reduced fare journeys were up by 9 million and season ticket journeys up by 4 million.

During the same period, receipts from full fares went up from £74·0 million to £79·7 million, reduced fares went up from £46·1 million to £54·1 million and season ticket receipts increased by £5·9 million.

Draw up tables which will summarize the above information and will bring out the principal changes between the two years, and will compute derived statistics where appropriate. Write brief comments on the changes between the two years.   (IoT)

6

Accuracy and Error

SUPPOSE it is desired to estimate the percentage by weight of nitrogen in a fertilizer mixture. Samples may be taken from different parts of the mixture on different days, sent to different laboratories, and analysed by different analysts using different methods. In the first place, there will be slight variations in the composition of the mixture from day to day and during the day, owing to changes in temperature, humidity and other factors. These variations are called sampling errors, but they are not errors in the usual sense of the word. In the second place, there will be errors of analysis, i.e. differences between laboratories, between methods used, and between individual analysts. Thirdly, there will be errors of measurement or observation.

There are thus several kinds of errors in experimental work, e.g. sampling errors, analytical errors and errors of observation, and they can never be completely eliminated. It should be noted that they are not necessarily mistakes. An error, in the statistical sense, is merely the difference between an estimate and the true or ideal value, which it may be impossible to determine exactly. It is these "errors" due to the natural variability of observational data that create the need for statistical methods.

Different types of data naturally have different degrees of variability. In the physical sciences variability is generally small, in biology and the social sciences it is relatively large, and in economics and commerce, where the statistician cannot experiment but must be content to observe, it is greater still, and the data themselves are subject to different kinds of errors from those met in experimental work. They may be broadly classified as errors of origin, errors of inadequacy and errors of manipulation. Errors of origin are due to defects in the data collected, e.g. false information caused by forgetfulness or misunderstanding on the part of the person supplying it (see page 21); errors of inadequacy arise when the items observed are too few to provide a representative sample of the population; and errors of manipulation are mistakes in counting or measuring, clerical errors, etc., in fact any mistakes made in handling the data.

Bias

If a quantity is such that its errors all tend to lie in the same direction, they are said to be biased. This does not mean that all errors are necessarily in the same direction, but there will be more in one direction than in the other, and the total of a number of items will almost certainly err on the side of the bias. For example, a table of incomes based on income-tax returns would show a downward bias because taxpayers are more likely to understate their incomes than to overstate them.

If errors on both sides are equally likely, they will tend to neutralize one another and are said to be unbiased or compensating errors. Thus, if a number of large sums of money are rounded off to the nearest thousand pounds, the errors are unbiased, varying between −£500 and +£500.

Approximation

This leads to the subject of approximation and precision. Not only is absolute accuracy generally impossible, but it frequently happens that such accuracy as can be obtained is unnecessary. The accountant must, of course, balance his accounts exactly, if only to guard against fraud, but it does not matter to the Managing Director whether the week's turnover was £35,485·63 or £35,486·32, and it may be sufficiently accurate to round off the amount to £35,500 or even £35,000. A fair estimate produced quickly may be much more useful than a precise figure three months late.

Very often all that is required is a rough estimate, and it is a waste of time to aim at needless precision. A slide rule will not normally give results correct to more than three significant figures, but it is an extremely useful device all the same. It is desirable, however, to show clearly what the precision of the figures presented really is. This can be done in various ways. Generally it is implied by the figures or tables themselves. Thus, if sums of money are shown as £256,000, £73,000 and so on, or better still if a column is headed "£000," it is clear that they have been rounded off to the nearest thousand pounds.

Another method is to show the range within which the true figure is believed to lie. Thus a population might be stated as

267,000±1,500, meaning that the estimate of 267,000 might be as much as 1,500 out either way. Another way would be to say that it lies between 265,500 and 268,500. Percentages are sometimes useful here: if a figure is believed to be correct within, say, 0·2 per cent, it can be shown as 463,000±0·2 per cent.

In arithmetical computations, however, it is a mistake to approximate too soon, especially if the final result is the difference of two nearly equal numbers. For example,

$$3572·45 - 3558·61 = 13·84$$

Clearly it would be wrong to round off to the nearest integer during the calculation and say $3572 - 3559 = 13$. It is always safest to retain two or three more decimal places or significant figures during the calculation than will be required in the answer.

A common difficulty is that if a column of figures is rounded off and then added up, the result may not agree with the true total. Consider the following—

$$
\begin{array}{r}
19,247 \\
31,473 \\
67,285 \\
\hline
118,005 \\
\end{array}
$$

If the three items are rounded off to the nearest thousand, the total becomes 117,000, but if the total is important, it should also be shown correct to the nearest thousand. The right way is not to "force" the addition as in "A" below, but to round off all items (including the total) independently as in "B."

A. The wrong way	B. The right way
Thousands	*Thousands*
19	19
32	31
67	67
118	118

Spurious Accuracy

In published statistics, such as trade returns, figures are often printed as recorded, not rounded off, even when they are subject to substantial revision. The reader is supposed to use his own discretion about rounding off or using them in further calculations. Care should be taken, however, to avoid giving the impression that the figures are more accurate than they really are. Thus it is legitimate to say "the population of Slowe-on-the-Uptake as recorded in the last Census was 5,267," or to say a few years after the Census that the population is "about 5,000." It is not correct to say, years after the Census, "the population is 5,267."

Foreign exchange is a common source of spurious accuracy. For example, it may be estimated that a fire in New York has caused damage to the value of a million dollars. When translated into British money, this might appear as £416,667, whereas "about £400,000" would be both safer and less misleading. The observant reader will detect many similar examples of spurious accuracy, and should scrupulously avoid adding to them.

Absolute and Relative Errors

The *absolute error* of an estimate is the actual difference between the estimate and the true value (which may be unknown), counted as positive or negative according as the estimate is greater or less than the true value. If x represents the estimate and X the true value, the error e is given by—

$$e = x - X$$

The *relative error* is the absolute error divided by the true value, i.e. e/X. If X is not known exactly, an estimate of it (not necessarily x itself, but perhaps an average of several observations) can be used without increasing the error unduly. Relative errors are often expressed as percentage errors. For example, if 26,757 is rounded off to 27,000, the absolute error is -243 and the relative error is—

$$\frac{-243}{26,757} = -0.91 \text{ per cent.}$$

Combination of Errors

The mathematical treatment of errors will be developed in later chapters, but the effect of errors in quantities subject to simple operations of arithmetic can be discussed briefly here. For simplicity, only two quantities will be considered, but the results can be generalized to any number.

Let x_1 and x_2 be the true values (assumed known) and e_1 and e_2 the errors, so that the observed values are $x_1 + e_1$ and $x_2 + e_2$. Then

$$(x_1 + e_1) + (x_2 + e_2) = (x_1 + x_2) + (e_1 + e_2) \tag{6.1}$$

and

$$(x_1 + e_1) - (x_2 + e_2) = (x_1 - x_2) + (e_1 - e_2) \tag{6.2}$$

These equations show, if common sense does not, that in adding or subtracting quantities liable to error, the errors themselves must be added or subtracted in the same way to arrive at the error in the result.

The maximum possible errors in sums or differences can easily be found from this rule. Thus, if two amounts, £34,000 and £1,400, are added together rounded off to the nearest £1,000 and £100 respectively, the possible errors are £500 and £50, and the possible error in the result is £550. If the second item had been subtracted from the first, the possible error would still have been £550, not £450, since the two errors might be in opposite directions, e.g.

$$(34,000 - 500) - (1,400 + 50) = 32,600 - 550$$

As the reader will see, the evaluation of possible errors in addition or subtraction only requires a little careful thought.

In multiplication and division it is the relative errors, not the absolute errors, that must be added or subtracted, since

$$(x_1 + e_1)(x_2 + e_2) = x_1 x_2 \left(1 + \frac{e_1}{x_2}\right)\left(1 + \frac{e_2}{x_2}\right)$$

$$\simeq x_1 x_2 \left(1 + \frac{e_1}{x_1} + \frac{e_2}{x_2}\right) \qquad (6.3)$$

provided the product

$$\frac{e_1 e_2}{x_1 x_2}$$

can be neglected. (The sign $\simeq$ denotes approximate equality.) Similarly, it can be shown that

$$(x_1 + e_1) \div (x_2 + e_2) \simeq \frac{x_1}{x_2}\left(1 + \frac{e_1}{x_1} - \frac{e_2}{x_2}\right) \qquad (6.4)$$

It must be remembered that in all the above formulae e_1 and e_2 may be either positive or negative. In calculating possible, as distinct from actual, errors, the magnitudes of the possible errors must be added together as if they were all positive. Thus, if an operator performs the operation $a \times b \div c$ on a slide rule and reads each time to an accuracy of 0·4 per cent, the result will be accurate within 1·2 per cent. It is, of course, extremely unlikely that he would make the maximum error each time, but that is another matter.

Exercises

6.1 Explain the terms

    (*a*) Absolute error.
    (*b*) Relative error.
    (*c*) Percentage error.

The three dimensions of a rectangular block of stone are 1, 2 and 3 feet respectively, with a relative error of 0·01 in each case. What is the greatest possible percentage error in the volume stated as 6 cubic feet, and what is the maximum absolute error? (IoS)

6.2 $S = P + Q$: S is estimated at 156,000 correct to the nearest 1,000, and P at 49,500 correct to the nearest 100. Find the value of S/Q with the possible limits of error. (IoS)

6.3 You have to report some figures which are accurate only within certain limits. Give examples showing *three* ways of indicating this fact for one of the figures. Why do such margins of error arise in figures?

The following figures are accurate to the nearest unit only. Work out the limits of error.

$$\frac{\left(19 - \frac{17}{4}\right)3}{11}$$

                                        (ICWA)

6.4 For three different areas, the number of households and the average consumption of coal per household are given below.

District	No of households	Tons/household
A	435,000	1·56
B	273,000	1·29
C	527,000	2·17

Make an estimate of the total consumption (*a*) for each separate district and (*b*) for the three districts combined, bearing in mind that the number of households is given to the nearest thousand and the average consumption to the nearest hundredth of a ton. (IoS)

6.5 Explain the method by which statistical figures are rounded and the reasons for doing so.

Round the commodity export values on the following table to the nearest £1,000 and indicate the errors of approximation in the rounded figures.

Commodity	Original values
	£
Spirits	11,361,617
Beer	1,142,896
Fruit juice and table waters	223,490
Cocoa preparations:	
With sugar	547,527
Without sugar	264,771
All other items	283,561
Total	13,823,862

(ACCA)

6.6 A motorist whose car has a broken distance recorder and who, therefore, measures his distances by map-reading, wishes to assess his
petrol consumption on a Continental trip. His petrol gauge indicates
his fuel consumption in litres with an error of ±10 per cent. He also
uses approximate conversion factors of 1 gallon = 4·5 litres and
1 mile = 1·6 Kilometres, instead of the more accurate 1 gallon = 4·55
litres and 1 mile = 1·61 Kilometres. Estimate his probable petrol
consumption in miles per gallon, giving limits of error, if he claims
to be getting 31 miles per gallon after a trip of exactly 615 miles.

(IoS)

7

Charts

THE great advantage of charts is that the salient features stand out immediately. They add nothing to the information contained in the statistics they illustrate, but they bring out clearly comparisons and trends which the figures themselves might only reveal under close scrutiny. There are, however, certain limitations. The number of items depicted must be small, or confusion will result, and there must be enough movement to give sufficient contrasts. Also, charts require considerable skill in drawing and are expensive to reproduce, especially in more than one colour.

Types of Charts

Most readers will be familiar with the commoner types of chart, particularly with *graphs* in which continuous lines or curves depict relations between two variables (generally denoted by x and y), represented on the graph by distances from two lines (the x- and y-axes) at right angles to each other. The student who needs to refresh his memory on the subject should refer to any elementary textbook on algebra.

Current usage does not distinguish clearly between the terms *diagram*, *figure* and *chart*. For convenience, the word *diagram* will be used to denote the simpler types of chart, generally involving lines, rectangles or circles. The words *chart* and *figure* can be used indifferently of both graphs and diagrams.

Most of the charts occurring in statistics are either graphs or similar to graphs in representing relations between two variables. For example, there is the time-series chart, showing how one variable, such as output, sales, population or prices, varies with time. More will be said about this later in the present chapter.

There is the frequency diagram, to be discussed in detail in Chapter 9, showing how frequency of occurrence varies with the value of some numerical characteristic. There is the scatter diagram, showing how one quantity depends partly on the value of another, e.g. a man's height and weight; this will be discussed in Chapter 14. There are also simpler diagrams, now to be described, generally designed for a wider public.

Pie Charts

The "pie-chart" is a popular device for showing how an aggregate is divided into its principal components. Fig. 7.1 shows how the turnover of a company might be divided into cost of materials, wages and salaries, other costs, and profit.

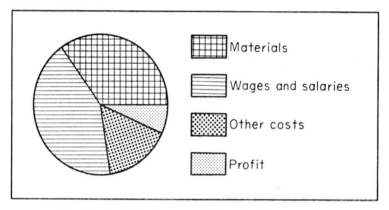

Fig. 7.1   Analysis of turnover of Dunn, Wright and Co Ltd

In this diagram, the various items are proportional to the areas representing them in the circle, or to the angles of the various sectors. They can be converted into percentages by dividing the angles at the centre by 360 and multiplying by 100. If it is desired to represent absolute magnitudes, other methods are preferable, e.g. bar charts.

Bar Charts

In the simplest form of bar chart, several items are shown graphically by horizontal or vertical bars of uniform width with lengths proportional to the values they represent. Fig. 7.2 shows a simple example of this.

If preferred, the actual figures can be omitted, and the countries can be given in a key similar to that shown in Fig. 7.1, the bars being shaded or coloured correspondingly.

A rather more complicated chart, combined with a table, is given in Fig. 7.3(a), representing the coal output of the three main producing countries in different years.

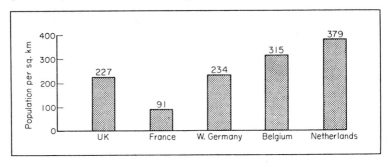

Fig. 7.2　Density of population in various countries in 1968

An alternative method of presenting the same data is shown in Fig. 7.3(b). Whereas Fig. 7.3(a) brings out the increase or decrease in each country's output, Fig. 7.3(b) focuses attention on the relative importance of the three countries at different times. It is irrelevant

Country	Year	Million tonnes	Million tonnes 100　200　300　400　500
UK	1953	228	
	1960	197	
	1967	175	
USA	1953	440	
	1960	392	
	1967	481	
USSR	1953	218	
	1960	356	
	1967	414	

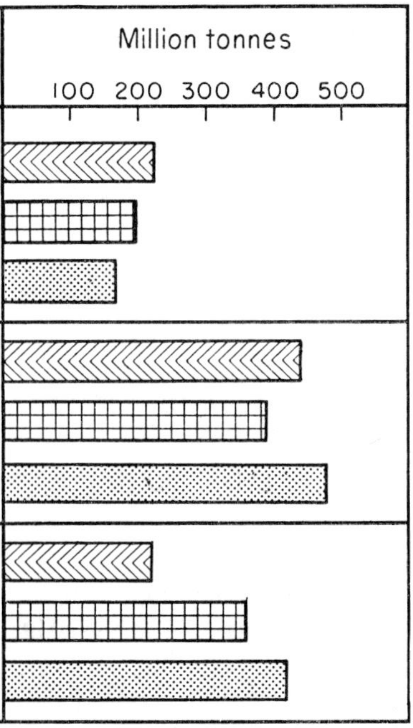

Fig. 7.3(a)　Coal output of chief producing countries in 1953, 1960 and 1967

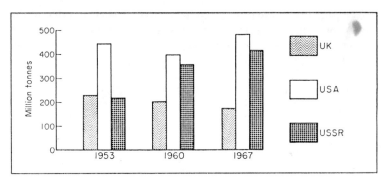

Fig. 7.3(b)   Coal output of chief producing countries in 1953, 1960 and
1967 (alternative form)

for this purpose whether the bars are horizontal or vertical, how
the figures are shown (if at all), whether the bars are contiguous or
separated, and whether the countries are distinguished by different
shading or, better still if possible, by different colours. All these
are largely matters of taste and money.

When it is desired to show an aggregate and its division into
components, the bars are drawn in length to the totals
and divided in the ratios of their components, as in Fig. 7.4.

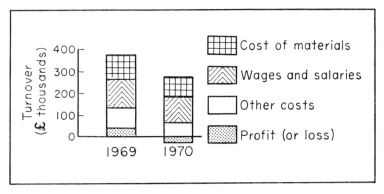

Fig. 7.4   Analysis of the turnover of Sherlock's Salted Milk Ltd,
1969 and 1970

A loss appears in this diagram as a negative profit below the base
line. This is included in the section representing "other costs," but
it is difficult to make this clear in the diagram. If all that is required
is the division of the total into percentages this can be shown by a
percentage bar chart as in Fig. 7.5.

The principle of the percentage bar chart is the same as that of the "pie chart," and there is little to choose between them except that on the former the actual percentages are more easily shown and read off.

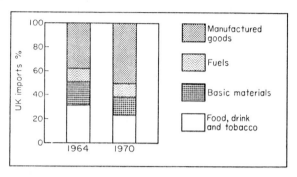

Fig. 7.5   Percentage analysis of UK imports by value, 1964 and 1969

Graphs of Time Series: Line Charts

The commonest type of chart in economic and commercial statistics is that of the time series, showing the progress of one or more quantities at successive times, usually at regular intervals. There are in fact, although the distinction is often blurred, two kinds of time series. In one, the quantity studied is one which has a certain value at any given moment, e.g. population or stocks; in the other, it is a total or average relating to periods of time, e.g. annual or monthly production, average prices or weekly earnings. In practice the distinction is not very important, since both kinds of series may be illustrated by the same kinds of charts.

Examples of time series have already been hinted at in Figs. 7.3(b) and 7.4, except that two years, or three at regular intervals, can hardly be said to constitute a series. If the data for these diagrams were continued for several years, either would provide a good example of a time series illustrated by a bar chart. It is simpler and usually more convenient, however, to use a line chart, in which points representing the data are joined by lines to aid the eye. Fig. 7.6 shows such a chart for production of cars in the United Kingdom from 1964 to 1969 with an earlier year (1956) for comparison. Notice that 1956, being so far removed in time from 1964, is joined to it by a broken line, whereas 1964 to 1970 are equally spaced horizontally and joined by continuous lines.

A rather more complicated example is given in Table 7.1 and Fig. 7.7. Here it was desired to show the annual value of imports into the United Kingdom from 1964 to 1969 and monthly values

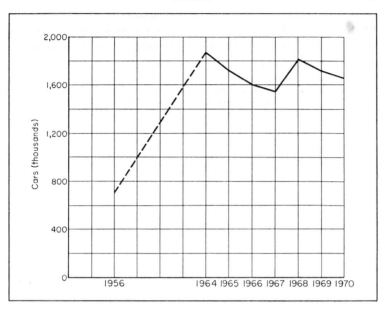

Fig. 7.6 UK production of cars 1956 and 1964–1970

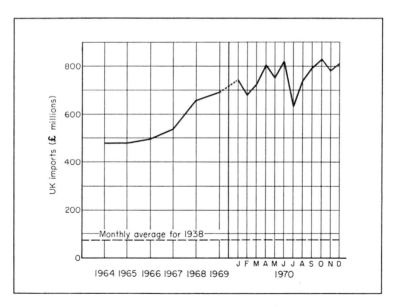

Fig. 7.7 Imports into the UK, 1938 and 1964–1970 (monthly averages or calendar months) (from Table 7.1)

thereafter, with a pre-war year (1938 this time) for comparison. In a case like this, where annual and monthly data appear together, it is necessary to give the annual data as monthly averages, and this should always be made clear.

Table 7.1   Imports into the UK, 1938 and 1964–1970
(monthly averages or calendar months)

	£ million			£ million
1938	77	January	1970	743
1964	475	February	,,	680
1965	479	March	,,	726
1966	496	April	,,	808
1967	536	May	,,	749
1968	658	June	,,	819
1969	693	July	,,	632
1970	754	August	,,	737
		September	,,	796
		October	,,	823
		November	,,	791
		December	,,	802

Opinions will differ about the best way to present these data. The aim in Fig. 7.7 has been to suggest methods not shown earlier. The 1938 average appears as a broken horizontal line, and the change of scale necessitated by monthly data is marked by a vertical line and a broken line from the 1969 point to that for January 1970. If, however, it is intended eventually to show the

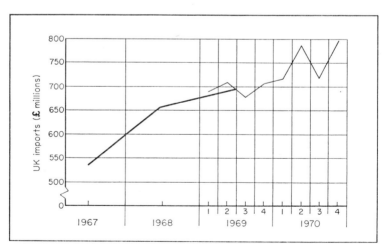

Fig. 7.8 Imports into the UK, 1967–1970 (monthly averages) (from Table 7.2)

average for 1970, the scale should be uniform throughout, even if the months are cramped together.

Fig. 7.8 shows how annual and quarterly data can be presented together. To make them comparable, they are all given as monthly averages. The lines joining quarterly points are lighter than those joining annual points, and the average for 1969 is shown midway between the second and third quarters.

Table 7.2   Imports into the UK, 1967–1970
(monthly averages)

		£ million
1967		536
1968		658
1969:	1st qr	687
	2nd qr	706
	3rd qr	677
	4th qr	705
	Average	694
1970:	1st qr	715
	2nd qr	785
	3rd qr	719
	4th qr	797
	Average	754

Broken Scales

The reader will notice another new feature in Fig. 7.8, viz. the broken vertical scale. Normally it is desirable for the scale to begin at zero and proceed uniformly, but sometimes the variation in the figures is small compared with their general level and would not show very well if this rule were strictly adhered to. It is then permissible to show a broken scale as in Fig. 7.8. This should never be done with a bar chart.

Two or More Variables

Two or three variables can easily be exhibited on the same chart if they are expressed in the same units. Thus exports could be shown on Fig. 7.7 by a second line distinguished from that for imports by being heavier or discontinuous. If coloured lines are used, four or five variables may be shown, e.g. sales of different commodities, but more than that would make the chart very confusing, especially if the lines kept crossing.

Variables in different units create further complications. If necessary, two different scales can be shown, one on the left and one on the right as in Fig. 7.9, the choice of relative scales being a matter for individual judgment. If colours can be used, each scale can be shown in the same colour as the corresponding variable.

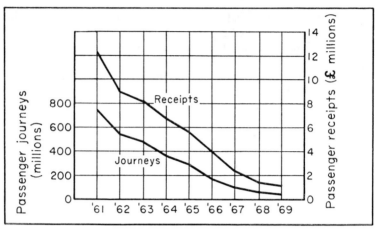

Fig. 7.9 Passenger transport on trolley-buses in the UK, 1961–1969
(from Table 7.3)

Table 7.3 Passenger transport on trolley-buses in UK, 1961–1969

Year	Passenger journeys (millions)	Passenger receipts (£ millions)
1961	756	12·2
1962	557	9·0
1963	476	8·1
1964	368	6·7
1965	285	5·7
1966	188	4·0
1967	106	2·4
1968	68	1·7
1969	50	1·3

The figure will look better and be easier to read if the scales are chosen so that one is a simple multiple of the other (in this case, 100). Alternatively, ratio scales may be used (see below).

Surface and Strata Charts

When the space between the line or curve and the base line is filled in with some kind of colour or shading, the presentation is often called a *surface chart*. Sometimes, when it is desired to show how an aggregate is divided into components, there are several layers in different colours or shadings, as in Fig. 7.10. This form of chart is called a *strata chart*.

The chief disadvantage of strata charts is that it is not easy to read off the magnitudes of the upper layers.

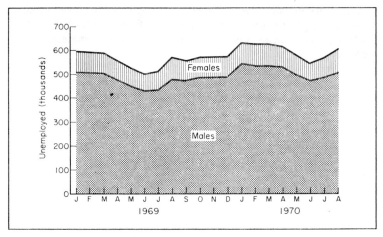

Fig. 7.10   Registered unemployed in Great Britain, 1969 and 1970

Ratio or Logarithmic Charts

For some purposes it is better to use the scale obtained by plotting log y instead of y. The result is known as a Ratio Scale or Logarithmic Scale. Readers familiar with the theory of logarithms will realize that the effect of the change is to compress the grid in the upper part of the graph and expand it in the lower part, so that equal vertical intervals as measured on the graph now represent not equal absolute changes, but equal percentage changes. Thus a rise from 1 to 2, from 3 to 6, or from 500 to 1,000 corresponds to the same vertical interval on the chart. It will be evident that the scale cannot start at zero, as a uniform scale normally does. If it starts at 1, the vertical distance of any point from the base line is proportional to the logarithm of the number it represents; hence the name.

This method is extremely useful when it is desired to study relative rather than absolute movements, when a series contains both very small and very large values, or when the series is one that might be expected to increase roughly in geometric progression, like population. A series increasing or decreasing in this way, e.g. a sum of money accumulating at a constant rate of compound interest, is represented on a logarithmic chart by a straight line.

Fig. 7.11 shows the data of Table 7.3 and Fig. 7.9 plotted on a logarithmic chart (or strictly speaking, on a semi-logarithmic chart, for only the vertical scales are logarithmic, time being shown on a uniform horizontal scale). As with most ordinary line charts, the sole purpose of the lines is to aid the eye. This graph shows the gap between the two lines gradually increasing, whereas in Fig. 7.9, because both journeys and receipts are falling sharply, the gap is also reduced. Fig. 7.11 shows that receipts have fallen relatively less than journeys, which means (as we all know) that the average fare has increased.

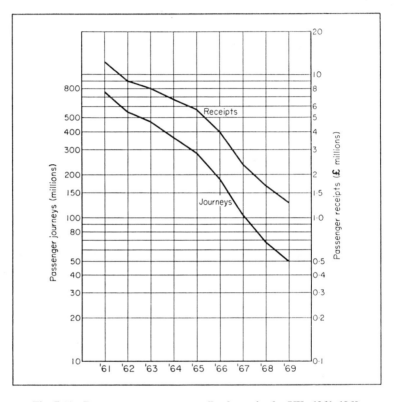

Fig. 7.11   Passenger transport on trolley-buses in the UK, 1961–1969
(semi-logarithmic chart) (from Table 7.3)

Special graph paper, known as semi-logarithmic paper, is available for these charts. It is scaled and ruled on the same principle as Fig. 7.11 and commonly gives two or three "cycles" of 1 to 10, e.g. "three-cycle" paper goes from 1 to 1,000. If such paper is not to hand, a logarithmic chart can be drawn by plotting the logarithms of the values on ordinary graph paper.

The chief characteristics of natural scale and ratio scale graphs can be summarized as follows—

Comparison of Natural Scale and Ratio Scale (Logarithmic) Graphs

Natural scale	Ratio scale
1. Equal vertical distances represent equal *absolute* changes. Lines of equal slope represent equal rates of change. A straight line denotes a continuous increase at simple interest.	Equal vertical distances represent equal *proportional* changes. Lines of equal slope represent equal *proportional* rates of change. A straight line denotes a continuous increase at compound interest.
2. Suitable for analysing an aggregate into its constituents.	Not suitable.
3. Zero and negative values may be shown.	Zero and negative values cannot be shown.
4. Must be definitely located with reference to a base line, whether shown on the graph or not.	No base line. The whole curve may be moved up and down without affecting its properties.
5. Not suitable when the *y* variable shows a great range of variation.	Eminently suitable in such cases.

Exercises

7.1 Illustrate the following data by means of bar charts *or* pie charts. Give the reasons for your choice of chart and state the relative advantages and disadvantages of the two methods of presentation.

Pens and pencils (manufacturers' sales) (£ thousand)			
	1956	1961	1966
Fountain pens	2,973	2,597	3,236
Ball point pens	2,361	3,405	4,535
Felt tip pens	—	—	677
Fibre tip pens	—	—	566
Pencils	1,788	2,013	1,985

Source: Board of Trade

(ICWA)

7.2 As an accountant you have been asked by your Board of Directors
to comment on the relative performance of the five firms making up
the industry. The only available information is as follows:

	Quantity sales		Turnover £ (000's)	
	1963	1964	1963	1964
Firm A	50,000	60,000	1,600	1,860
,, B	40,000	56,000	1,240	1,680
,, C	10,000	15,000	300	450
,, D	30,000	30,000	960	930
,, E	20,000	26,000	640	806

Assume all quantity sales are comparable.

Comment suitably to your Board, and in replying, any graphs or
charts need not be accurate but must indicate salient features.

(ACCA)

7.3 Draw charts to bring out the main features shown by the following
figures of passenger movement. Say *briefly* what these main features
are.

Destination of passengers leaving the UK

	Thousands				
	1961	1962	1963	1964	1965
By sea Europe and Mediterranean					
countries	2,486	2,631	2,915	3,024	3,295
North America	121	128	127	115	108
Rest of world	159	159	169	169	154
By air Europe and Mediterranean					
countries	2,477	2,716	3,030	3,400	3,862
North America	398	459	528	670	782
Rest of world	164	183	210	260	322

Source: Annual Abstract of Statistics   (IoT)

7.4 *Forecast car registrations over the period March 1966 to February 1967*

Month	Actual registrations (thousands)	Predicted registrations (thousands)
March 1966	137·2	135·4
April	137·1	115·4
May	120·3	100·0
June	102·7	99·7
July	99·7	95·5
August	59·4	57·6
September	54·6	59·3
October	53·0	56·7
November	52·8	49·6
December	39·3	46·4
January 1967	103·6	89·3
February	92·9	92·9

(*Source: The Statistician*, Vol. 17, No. 3, 1967)

Plot these two sets of figures on a graph and comment on the result.   (IoS)

7.5 The annual production of the three departments of the Lester Manufacturing Company from 1955 to 1965 is shown in the following table:

Annual Production Lester Manufacturing Company 1955–1965

Year	Thousands of units		
	Department A	Department B	Department C
1955	150	190	160
1956	170	230	170
1957	200	150	200
1958	240	210	150
1959	200	280	220
1960	250	300	100
1961	270	230	200
1962	300	220	260
1963	280	320	200
1964	350	280	270
1965	400	250	150

(*a*) Draw:

    (i) a line chart or band chart to show the information in the table,

    (ii) a bar chart to show the facts for the years 1955 and 1965,

    (iii) a pie chart to show the facts for 1965.

(*b*) What are the merits of presenting statistical information in graphic form?   (ACCA)

7.6 Draw a chart to illustrate changes in coal dispatches and the tonnage moved by rail and say what this tells you about the rail share of the traffic. Draw a second chart comparing the rail tonnage and ton-mileage. Discuss what this shows.

Year	Coal dispatches	British Railways: Coal class traffic	
	Million tons	Million tons	Thousand million ton miles
1953	196	175	10·7
1954	202	173	10·5
1955	203	166	10·2
1956	202	168	10·2
1957	197	167	9·9
1958	185	153	8·9
1959	174	144	8·0
1960	183	148	8·1
1961	181	146	7·7
1962	181	145	7·3

Sources: Ministry of Power Statistical Digest;
British Transport Commission Report and Accounts

(IoT)

7.7

Year	Production of goods vehicles			
	Number of vehicles with carrying capacity of:			
	under 15 cwt	15 cwt to 6 tons	over 6 tons	Total
1954	143,384	97,020	20,229	260,633
1955	188,385	117,194	24,272	329,851
1956	155,540	105,553	25,353	286,446
1957	142,148	109,182	24,018	275,348
1958	166,027	101,501	30,257	297,785
1959	185,029	119,169	48,907	353,105
1960	221,180	152,851	64,893	438,924
1961	246,705	128,980	67,432	443,117
1962	215,677	134,544	58,293	408,514

Sources: Board of Trade, Ministry of Supply;
Ministry of Transport and Civil Aviation

Study the above table carefully and illustrate its salient features by means of a graph, calculating where necessary any secondary statistics.

(ICWA)

7.8 From the table below, draw a diagram to illustrate the changes that have taken place in the age distribution of the population of Great Britain since 1851.

Age distribution per 1,000 persons			
Year	0–14	15–64	65 and over
1851	355	598	47
1891	351	601	48
1911	308	639	53
1939	214	697	89
1947	215	681	105
1957	229	653	118

Source: Stern, Britain Yesterday and To-day

(IoS)

7.9 Draw charts to bring out the main features of the statistics contained in the following table, in particular comparing the figures at current prices with those at constant prices. Comment on the changes in the pattern of consumer expenditure revealed by your charts.

Consumers' expenditure on transport and vehicles

	£ million				
	1963	1964	1965	1966	1967
At current prices cars and motor cycles— purchases	733	843	799	807	898
running costs	666	775	928	1,058	1,161
All other travel	663	801	736	778	809
At constant (1958) prices cars and motor cycles— purchases	891	1,032	976	975	1,061
running costs	608	699	773	845	910
All other travel	546	563	570	589	597

Source: National Income and Expenditure, 1968

(IoT)

7.10 "In 1912 there were 250,000 road vehicles in Britain. Twenty years later there were ten times as many. By 1952 the number had doubled again to double yet again by 1962 when there were 10½ millions. There is now the prospect of 18 million vehicles by 1970, 25 millions by 1980 and more than 30 millions by the end of the century." (*The Guardian*, 7 November 1966.)

Draw a graph to illustrate this statement and comment on the shape of the graph.   (IoS)

7.11 The sales figures given below relate to the same commodity:

	Total UK sales (thousands)	Sales by company A
1954	91	646
1955	95	672
1956	99	704
1957	105	747
1958	111	823
1959	115	902
1960	122	1,084
1961	127	1,305
1962	133	1,561
1963	140	2,010
1964	147	2,628
1965	154	3,475

Plot both sets of figures on the same log graph and comment on your findings.   (ICWA)

8

Analysis of Time Series

THIS chapter deals with elementary methods of analysing time series. A time series is constructed by plotting successive values of some variable (x) at successive intervals of time (t). It is implied that this series is long enough to enable valid statistical inferences to be drawn. The most interesting series are those of "economic indicators," i.e. those which measure aspects of current economic activity as represented by statistics of prices, production, transport, bank clearings and so on.

In general, x is a composite quantity: it represents the result of a number of influences which cannot be isolated for purposes of study and measurement. Nevertheless, statistical methods provide a means of estimating the contributions of the more important factors, and this is the problem to be considered.

According to current practice, a time series may be regarded as involving as many as five components, namely—

1. The *Trend*, or the course the series would take over a long period in the absence of disturbing factors.

2. *Cyclical fluctuations*, or wave-like disturbances corresponding, for example, with the movements of the Trade Cycle or the stock market.

3. *Seasonal variations* associated with the harvests, the weather, Christmas and other festivals, and possibly the varying length of the month, although this is of minor importance.

4. *Catastrophic movements* caused by unusual or unexpected events, e.g. the Suez crisis of 1956 or the severe winter of 1963.

5. *Residual variations*, which include all movements not already covered in headings 1 to 4.

For elementary purposes it will be sufficient to consider three components, namely: the trend (1), seasonal variations (3), and deviations from the trend—the result of pooling (2), (4) and (5)—and the rest of this chapter has been written on these lines.

Undoubtedly there are seasonal effects in many time series, although it is generally difficult to estimate them numerically. It is also, as a rule, possible to detect a general trend over a period, although the short-term trend may be quite different from the long-term trend. There is no reason why the above analysis should not be used to explain current events, but to use it for forecasting is hazardous because much depends on politics or unforeseen circumstances. If, however, there is a well-defined trend over a long period, it may be reasonable to extrapolate into the not-too-distant future.

Finding the Trend

Various methods have been devised for breaking up series into the elements mentioned above. It is generally most important to find the trend, or perhaps it would be better to say *a* trend, as there is seldom an exact and unique solution to this problem.

It is always helpful to plot the series on a graph, either on a natural scale or on a semi-logarithmic scale. The latter will be better if there is a tendency for the series to increase or decrease in geometric progression, as the trend will be more like a straight line on a logarithmic chart than on uniform graph paper. Assuming that, in the long run, deviations from the trend will be as often positive as negative, and vice versa, the trend can be estimated by any of the following methods:

1. Drawing a smooth freehand curve so that about the same number of points lie on either side of it. This method demands greater skill and judgment than are possessed by the average beginner, and even trained operators may differ considerably in their results.

2. Fitting a straight line or a mathematical curve to the points by methods to be described in Chapter 14.

3. Calculating a suitable *Moving Average* (see below).

Moving Average

This device consists in substituting for the original series a smoother series, each term of which is the average of a number (always the same number) of terms of the original series. Thus, if the original series is—

    81, 92, 100, 106, 116, 127, etc.

and a 5-point moving average is required, the first term will be—

$$\tfrac{1}{5}(81+92+100+106+116)=99$$

and the second—

$$\tfrac{1}{5}(92+100+106+116+127)=108 \cdot 2,$$

and so on. At each step, one term of the original series is dropped and another is introduced. In fact, the second average could be obtained simply by adding $\tfrac{1}{5}(127-81)=9 \cdot 2$ to the first. Table 8.1 shows how a 5-point moving average is calculated for a series of weekly data, and Fig. 8.1 shows the result plotted on a graph.

Table 8.1 Percentage time lost through sickness in a factory over a period of 21 weeks

Week	Percentage time lost	Sum of 5	Moving average
1	2·5		
2	3·3		
3	4·0		4·1
4	4·7		4·6
5	6·0	20·5	4·7
6	5·0	23·0	4·6
7	3·6	23·3	5·2
8	3·7	23·0	5·6
9	7·8	26·1	5·5
10	7·7	27·8	5·4
11	4·7	27·5	5·3
12	3·0	26·9	4·1
13	3·2	26·4	3·3
14	2·1	20·7	2·5
15	3·3	16·3	2·0
16	1·1	12·7	1·5
17	0·4	10·1	1·3
18	0·7	7·6	1·1
19	0·8	6·3	1·3
20	2·4	5·4	
21	2·4	6·7	

Each sum of five items after the first is obtained by adding to the previous sum of five the difference between the new term and the one dropped. Thus, for weeks 11–15, the sum is—

$$20 \cdot 7+(3 \cdot 3-7 \cdot 7)=16 \cdot 3$$

The final entry in the third column should be checked by direct addition, and if the series is a long one, a few intermediate checks should be provided.

Notice that each average is plotted in the centre of the period it covers. This is clearly right, for in the extreme case of a series in perfect arithmetic progression, the moving average "centred" in this way would coincide with the original series, and both would

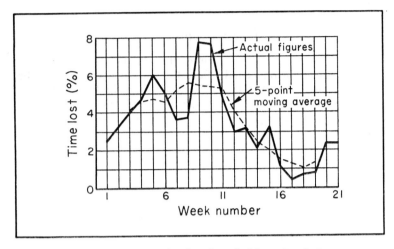

Fig. 8.1   Percentage time lost through sickness in a factory
(from Table 8.1)

be depicted by the same straight line. It means, however, that the moving average will miss nearly half the period (two years in the case of a 5-year average) at each end.

As will be seen in Fig. 8.1, the moving average still shows kinks, although it is much smoother than the original series. It can, if desired, be further smoothed by a freehand curve or by a further averaging process.

Where there are cyclical fluctuations, the number of terms in the moving average will generally depend on the period of the cycle to be eliminated. Thus a 7-year cycle may be eliminated by a 7-year (or perhaps a 14- or 21-year) average, and seasonal movements by a moving average based on twelve months or four quarters. In the latter case, in fact whenever the moving average has an even number of terms, it can still be plotted in the middle of the period although it will now lie between the two middle items, not in the same vertical as a single middle item. There is no great objection to this, but for the purpose of trying to measure seasonal variations it is desirable for the moving average to relate to the same time or period as one of the original items. This can be arranged by averaging a step further. Suppose, for example, it is desired to plot the trend, and seasonal deviations from the trend, of a series of quarterly data. The average of the first four quarters will be midway (in time) between the second and third, and the average of the second to fifth quarters midway between the third and fourth. Clearly, then, the average of those two averages will give the trend for the third quarter. It is most simply obtained by first calculating the sums of four quarters, or *moving annual totals*, then adding those in pairs and dividing by 8, as in Table 8.2. For monthly data the procedure would be similar, but each MAT would be the sum of twelve items and the divisor would be 24 instead of 8.

Table 8.2   Sales of a certain fertilizer by Gromore and Sons Ltd
(Thousand tons)

Year and quarter	Sales	Moving annual total	Sum of two MATs	Moving average	Deviation from trend
1966: 1	60				
2	65				
3	20			47	− 27
4	44	189		47	− 3
1967: 1	62	191	380	47	+ 15
2	58	184	375	49	+ 9
3	28	192	376	52	− 24
4	50	198	390	53	− 3
1968: 1	85	221	419	52	+ 33
2	42	205	426	52	− 10
3	33	210	415	55	− 22
4	44	204	414	63	− 19
1969: 1	118	237	441	65	+ 53
2	71	266	503	65	+ 6
3	20	253	519	66	− 46
4	58	267	520	66	− 8
1970: 1	110	259	526	68	+ 42
2	83	271	530	68	+ 15
3	22	273	544		
4	55	270	543		

Seasonal Variations

When a series is strongly seasonal, like sales of ice-cream or the number of houses built each month, the fluctuations are apt to be confusing, and various methods have been devised for removing them so that one month may legitimately be compared with another. The best and most straightforward way of doing this is shown in Tables 8.2, 8.3 and 8.4, with quarterly data for simplicity. Monthly data can be treated on the same lines. It must be stressed, however, that this procedure is only legitimate when seasonal movements are very marked and reasonably consistent from year to year.

The moving average is first calculated as already described, and subtracted from the corresponding actual value to give the deviation from trend, e.g. $20-47= -27$. These deviations are rearranged as in Table 8.2 and totalled for each month or quarter

Table 8.3 Fertilizer sales—seasonal movements
(Derived from Table 8.2)

Year	Quarter				Total
	1	2	3	4	
1966			− 27	− 3	
1967	+ 15	+ 9	− 24	− 3	
1968	+ 33	− 10	− 22	− 19	
1969	+ 53	+ 6	− 46	− 8	
1970	+ 42	+ 15			
Total	+ 143	+ 20	− 119	− 33	+ 11
Adjusted total	+ 140	+ 17	− 121	− 36	0
Average	+ 35	+ 4	− 30	− 9	0

Table 8.4 Fertilizer sales corrected for seasonal movements
(Thousand tons)

Year and quarter	Actual sales	Average seasonal movement	Sales (adjusted)
1966: 1	60	+ 35	25
2	65	+ 4	61
3	20	− 30	50
4	44	− 9	53
1967: 1	62	+ 35	27
2	58	+ 4	54
3	28	− 30	58
4	50	− 9	59
1968: 1	85	+ 35	50
2	42	+ 4	38
3	33	− 30	63
4	44	− 9	53
1969: 1	118	+ 35	83
2	71	+ 4	67
3	20	− 30	50
4	58	− 9	67
1970: 1	110	+ 35	75
2	83	+ 4	79
3	22	− 30	52
4	55	− 9	64

of the year. If the totals do not add up to zero, they should be adjusted as shown before the averages for each quarter are calculated. These averages give the best available estimates of the seasonal effects. Finally, these estimated seasonal fluctuations are subtracted from the original items as in Table 8.4 to obtain "corrected" sales, i.e. what the sales would have been without the seasonal effect.

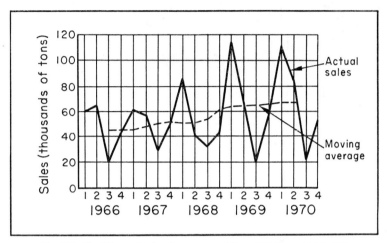

Fig. 8.2   Fertilizer sales by Gromore and Sons Ltd (from Table 8.2)

The results are shown graphically in Figs. 8.2, 8.3 and 8.4. As generally happens, the deviations from trend are partly seasonal and partly what, for want of a better term, is described as "random." The adjusted series is smoother than the original but not as

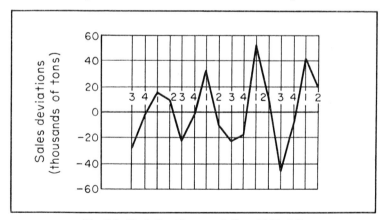

Fig. 8.3   Deviations from trend (from Table 8.2)

smooth as the trend line, because, although the seasonal effect has been removed (as far as possible), the residual or random fluctuation is left in. The movements do not show the same regularity as in the original series.

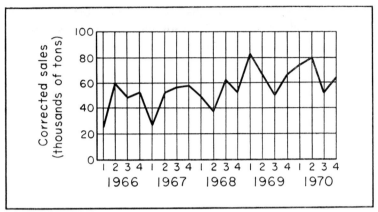

Fig. 8.4 Fertilizer sales, corrected for seasonal movements (from Table 8.4)

Table 8.5 shows the final analysis of the original data into trend or moving average, seasonal movement, and residual. Since the moving average cannot be calculated for the two quarters at each end of the original period, these quarters are omitted from the analysis.

Table 8.5 Analysis of fertilizer sales
(Thousand tons)
(Data of Table 8.2)

Year and quarter	Trend	Seasonal movement	Residual	Actual sales
1966: 3	47	− 30	+ 3	20
4	47	− 9	+ 6	44
1967: 1	47	+ 35	− 20	62
2	49	+ 4	+ 5	58
3	52	− 30	+ 6	28
4	53	− 9	+ 6	50
1968: 1	52	+ 5	− 2	85
2	52	+ 4	− 4	42
3	55	− 30	+ 8	33
4	63	− 9	− 10	44
1969: 1	65	+ 35	+ 18	118
2	65	+ 4	+ 2	71
3	66	− 30	− 16	20
4	66	− 9	+ 1	58
1970: 1	68	+ 5	+ 7	110
2	68	+ 4	+ 1	83

The method described above allows automatically for the varying length of the month and for most public holidays. A further adjustment may be necessary, however, if Easter occurs in April in one year and in March in another.

An Alternative Model

So far we have assumed that, if there were no random or residual variation, the seasonal differences from trend would be constant. This is an example of what, in the modern idiom (some would say jargon), is called a *statistical model*. A model is a hypothesis, often expressed in the form of equations or other mathematical terms, which we think may adequately describe the underlying processes. Usually it is over-simplified, but it often fits the facts well enough to enable us to use it for forecasting and similar purposes.

This particular model, however, may not be the best for a time series. If the general level of sales doubles over a period, it is likely that both seasonal and residual variations will, on the average, double also. In other words, if there is a definite trend upwards or downwards, it is unlikely that seasonal differences from trend will be constant; it is more reasonable to suppose that *seasonal factors* will be constant, i.e. that in the absence of residual variation, the ratio of any item to the corresponding trend value will be constant for any given month or quarter of the year.

Table 8.6   Sales by Gromore and Sons Ltd (alternative model)

Year and quarter	Sales (1,000 tons)	Moving average	Sales as percentage of moving average
1966: 1	60		
2	65		
3	20	47	43
4	44	47	94
1967: 1	62	47	132
2	58	49	118
3	28	52	54
4	50	53	94
1968: 1	85	52	163
2	42	52	81
3	33	55	60
4	44	63	70
1969: 1	118	65	182
2	71	65	109
3	20	66	30
4	58	66	88
1970: 1	110	68	162
2	83	68	122
3	22		
4	55		

Tables 8.6, 8.7 and 8.8 show how seasonal factors and seasonally corrected figures are obtained with this second model. First, each quarter's sales are expressed as a percentage of the moving average,

e.g. for the third quarter of 1968—

$$\frac{33}{55} \times 100 = 60$$

Next these percentages are averaged in Table 8.7, just as the differences were in Table 8.3. If necessary, the totals should be adjusted so that the average seasonal factors (expressed as percentages) add

Table 8.7   Calculation of seasonal factors

Year	Quarter				Total
	1	2	3	4	
1966			43	94	
1967	132	118	54	94	
1968	163	81	60	70	
1969	182	109	30	88	
1970	162	122			
Total	639	430	187	346	1,602
Average	160	107	47	86	400

Table 8.8   Sales by Gromore and Sons Ltd, seasonally corrected

Year and quarter	Actual sales (1,000 tons)	Seasonal factor	Adjusted sales (1,000 tons)
1966: 1	60	160	38
2	65	107	61
3	20	47	43
4	44	86	51
1967: 1	62	160	39
2	58	107	54
3	28	47	60
4	50	86	58
1968: 1	85	160	53
2	42	107	39
3	33	47	70
4	44	86	51
1969: 1	118	160	74
2	71	107	66
3	20	47	43
4	58	86	67
1970: 1	110	160	69
2	83	107	78
3	22	47	47
4	55	86	64

up to 400 exactly; in this example it happened to be unnecessary, allowing for some rounding off. Finally, the actual sales are divided by the appropriate factors to obtain seasonally corrected figures.

Thus, for the third quarter of 1968—

$$\frac{33}{0.47} = \frac{33 \times 100}{47} = 70$$

These seasonally corrected figures are shown in Fig. 8.5, which should be compared with Fig. 8.4. The residual variation is still large, but rather less than in Fig. 8.4, which suggests that the

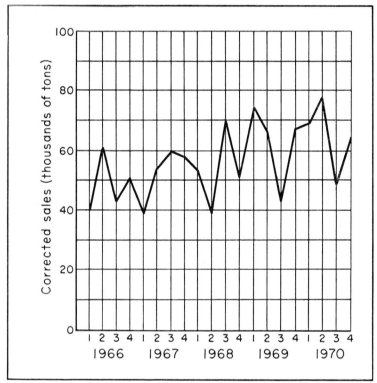

Fig. 8.5 **Fertilizer sales, corrected for seasonal movements (alternative model) (from Table 8.8)**

second model is better than the first, at least in this example. It may not always be; each case must be judged on its own merits. In practice, the second model is usually preferred, but the first model is nearly always used in examination questions for the same reason as the data are generally quarterly, not monthly: to keep the amount of calculation in one question within reasonable limits.

Z-Charts

The Z-chart is so called because its three component curves roughly form the letter Z, or rather a succession of Zs. It can be adapted to any period, but each Z normally takes one year, and it will be assumed in what follows that this is the period covered by the Z. The three components are:

1. The curve of the original data, generally but not always monthly.

2. The cumulative curve of the original data, e.g. the point for May 1969 will represent the total sales of, say, the five months January to May 1969.

3. The moving annual total, e.g. the point for May 1970 now represents the twelve months June 1969 to May 1970. The MAT is often extremely useful, quite apart from the Z-chart, as seasonal variations are automatically cancelled out.

Table 8.9   Sales of "Boxo," 1969–1971

Year and month	Monthly sales £000	Cumulative sales for year £000	Moving annual total sales £000
1969: January	45	45	
February	46	91	
March	53	144	
April	31	175	
May	48	223	
June	58	281	
July	49	330	
August	22	352	
September	82	434	
October	70	504	
November	65	569	
December	79	648	648
1970: January	62	62	665
February	59	121	678
March	76	197	701
April	44	241	714
May	54	295	720
June	76	371	738
July	54	425	743
August	31	456	752
September	68	524	738
October	76	600	744
November	83	683	762
December	104	787	787
1971: January	97	97	822
February	92	189	855
March	105	294	884

Clearly if the same scale is used for all three curves the cumulative curve will coincide with (1) at the beginning of the year and with (3) at the end of the year, and the Z will be one continuous curve, as in Fig. 8.6. Sometimes, however, the monthly data are plotted

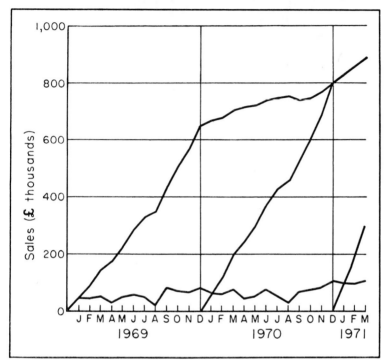

Fig. 8.6 Z-Chart representing sales of "Boxo" (from Table 8.9)

on a larger scale than the other two (say five times as large) to bring them out more clearly. This is certainly necessary with weekly or daily data.

Exercises

8.1 The figures overleaf were taken from the *Monthly Digest of Statistics:*

(*a*) Round the figures of the number of passenger journeys to the nearest 10 million.

(*b*) Calculate a 5-year moving average of the rounded series.

(*c*) Plot the figures on a graph.

(*d*) What is the maximum possible error due to rounding on each moving average?

British Railways	
Year	Total number of passenger journeys (millions)
1950	981·7
1951	1,001·3
1952	989·0
1953	985·3
1954	1,020·1
1955	993·9
1956	1,028·5
1957	1,101·2
1958	1,089·8
1959	1,068·8
1960	1,036·7
1961	1,025·0
1962	965·0
1963	938·0
1964	927·6
1965	865·1

Source: Ministry of Transport

(ICWA)

8.2 Compute the trend from the following data, stating how you determine the period of cyclical fluctuation.

Year	Sales per head (lb)	Year	Sales per head (lb)
1	11·5	12	14·0
2	11·2	13	12·8
3	11·9	14	13·0
4	13·2	15	13·7
5	13·9	16	15·1
6	13·4	17	16·2
7	11·8	18	15·9
8	11·7	19	14·6
9	12·4	20	14·5
10	13·6	21	15·3
11	14·9	22	16·9

(IoS)

8.3 Describe concisely the steps involved in calculating, by the method of moving averages, a seasonally-adjusted series. Of what use is such a series in a transport undertaking? By analysing his receipts in recent years, a transport operator estimates that on average his takings each January are 20 per cent below trend, those in February are 10 per cent below, and those in March 5 per cent below. In January,

February and March 1963, his actual takings were £1,000, £1,100 and £1,150 respectively. After allowing for seasonal influences, would you conclude that his receipts were rising or falling during this period?

(IoT)

8.4 The table below shows the sales of a certain article in each quarter over a period of 5 years. Using the method of moving averages, produce figures which show—

(a) The seasonal variation.
(b) The trend after eliminating seasonal variation.

Year	1	2	3	4	5
Spring	58	79	97	113	136
Summer	29	48	66	91	105
Autumn	49	68	85	100	123
Winter	89	107	134	148	170

(IoS)

8.5 Eliminate the seasonal variation from the following quarterly turnover figures of Mailorder Limited.

Year and quarter	Turnover £
1956: 1	3,372
2	4,106
3	4,388
4	3,744
1957: 1	3,214
2	4,166
3	4,624
4	4,142
1958: 1	3,356
2	4,282
3	4,470
4	3,696
1959: 1	3,112
2	3,990
3	4,222
4	3,608
1960: 1	2,980
2	3,746
3	3,906
4	3,394
1961: 1	2,918
2	3,898

Explain the purpose of removing the seasonal effects from the above data. (ACCA)

8.6 Deliveries of Woven Wool and Mixture Fabrics, excluding Blankets.

	Million square yards
1948	415
1949	439
1950	450
1951	418
1952	378
1953	412
1954	414
1955	410
1956	397
1957	394
1958	349
1959	365
1960	367
1961	352
1962	328
1963	325

Source: Annual Abstract of Statistics

From the above table

(a) calculate a 5-year moving average;

(b) represent the original series and the moving average on the same graph;

(c) by extrapolating the statistical trends, estimate the probable deliveries of woven wool and mixture fabrics for 1970.   (ICWA)

8.7 Use the method of moving averages to fit a trend line to the following quarterly figures of UK passenger movements by sea and air. Calculate seasonal adjustments. Draw a graph of the original data and the trend, and hence forecast UK passenger movements by sea and air for the 1st and 2nd quarters of 1968.

Year	1st quarter	2nd quarter	3rd quarter	4th quarter
1964	1,023	2,268	4,383	1,249
1965	1,075	2,615	4,843	1,367
1966	1,201	2,807	5,363	1,541
1967	1,411	2,938	5,613	1,708

(IoT)

9

Frequency Distributions

FREQUENCY distribution is one of the fundamental ideas of statistics. It has already been noted that before a large number of observations can be analysed, or even made readily intelligible, they must be sorted into a convenient number of groups, or classes. In the problems now to be discussed, the data are sorted according to the numerical value of some characteristic, called a *variable* (some writers prefer to call it a *variate*). Thus, a number of people might be sorted according to their height, weight, age, or any other characteristic capable of being measured. The number of items in any group is called the *frequency* of that group, and the resulting distribution is called (not unnaturally) the *frequency distribution*.

Table 9.1 shows a simple example of a frequency distribution, one of 500 families classified according to the number of children in a family.

Continuous and Discrete Variables

A variable that can take only discrete values, as in the above example, is said to be *discrete* or *discontinuous*. Obviously a man cannot have 2·63 children or pay 3·748p for a bus fare. Other examples of discrete variables are the number of houses in a street or rooms in a house, the number of apples bought in 1 lb, and the price of an ice-cream.

A *continuous* variable, on the other hand, can take any value whatever within a certain range—or, more probably, an uncertain range. Thus, a temperature need not be an exact number of degrees, but may theoretically be measured to several decimal places, and any object whose temperature is rising will, during the process,

75

take every possible temperature between the initial and final values. Similar examples are the speed of a vehicle and the height of a growing plant. In short, a continuous variable can take an infinite number of values.

Table 9.1   Frequency distribution of 500 families by number of children

No of children in family	No of families
0	179
1	205
2	78
3	26
4	7
5	3
6	1
7	–
8	1
Total	500

The distinction between continuous and discrete variables, however, tends to become blurred. In the first place, it is often desirable to treat discrete variables with many possible values as "almost continuous." For instance, suppose a large firm records the earnings of each worker in a given week; the variable may be any whole number of pence up to, say, 4,000. Strictly speaking, it is discrete, but the gap between successive values is so small and the number of possible values so large that the variable may be regarded for practical purposes as continuous. On the other hand, it is often necessary to treat continuous variables as discrete by rounding off values to the nearest inch, pound, etc. This can generally be done without appreciable loss of accuracy. As will be shown in this and subsequent chapters, it greatly facilitates work and simplifies the arithmetic.

Grouping the Data

In order to obtain a manageable frequency distribution when the variable is continuous, or exhibits a large number of discrete values, the range of possible values must be divided into a suitable number of sub-ranges and the data sorted accordingly. A group of men might be classified according to height and sorted into the following *groups*, or *classes*: under 5 ft; 5 ft and under 5 ft 2 in.; and so on (see Table 9.2).

Table 9.2 Frequency distribution of
1,000 men by height

Height (inches)	No of men
Under 58	2
58 and under 60	5
60 and under 62	14
62 and under 64	60
64 and under 66	187
66 and under 68	304
68 and under 70	263
70 and under 72	121
72 and under 74	36
74 and under 76	7
76 and over	1
Total	1,000

The number of items in a particular class is called the *class-frequency*, and the range of the variable in that class the *class-interval*. In Table 9.2 the class-frequencies are 2, 5, 14, etc., and the class-interval is 2 inches throughout, except at the extremes, where there is no definite class-interval.

It is usual to have a uniform class-interval throughout, as this gives a much clearer picture of the distribution than class-intervals of varying size. The reader can readily imagine what a confused impression Table 9.2 would have given if the men had been grouped as follows: under 5 ft; 5 ft and under 5 ft 3 in.; 5 ft 3 in. and under 5 ft 4 in.; and so on.

In some cases, however, it is impossible to keep the class-interval constant and at the same time to present a satisfactory frequency table. A good example is the distribution of incomes. A large class-interval would sacrifice all the detail about the smaller incomes, which form the great majority, and a small class-interval would either give an unwieldy table or involve lumping together all incomes over, say, £5,000.

Table 9.3, taken from the Blue Book on *National Income and Expenditure* for 1969, will illustrate the difficulty. The variable is income after tax, and there are eleven classes, with eight different class-intervals, including the open group "20,000 and over."

Table 9.3 Frequency distribution of incomes after tax in the
United Kingdom in 1967

Range of incomes after tax		Number of incomes (thousands)
Not under	Under	
£	£	
50	250	2,338
250	500	5,906
500	750	5,418
750	1,000	4,822
1,000	1,500	6,466
1,500	2,000	1,832
2,000	3,000	730
3,000	5,000	224
5,000	10,000	63
10,000	20,000	1
20,000		0
Total		27,800

Choice of Class-Intervals and Class-Limits

The student will learn from experience how to group the data in
any particular case, but it is generally best to have between ten
and twenty classes with equal class-intervals of convenient size.
Great care should be taken to define the class-limits, i.e. the end
values of each class. Thus, Table 9.2 might have been written in the
following way—

Table 9.4 Frequency distribution of
1,000 men by height
(How *NOT* to do it)

Height (inches)	No of men
–58	2
58–60	5
60–62	14
62–64	60
64–66	187
66–68	304
68–70	263
70–72	121
72–74	36
74–76	7
76–	1
Total	1,000

The table is ambiguous because it is not clear whether a man recorded as exactly 64 inches tall should be placed in the fourth or the fifth class, whereas Table 9.2 makes it clear that he is placed in the fifth. The critical student may object that no man would be exactly 64 inches tall, but that any man would be either under or over 64 inches if measured accurately enough. As already remarked however, it is not possible to measure with perfect accuracy, and in all probability the heights have been taken to the nearest $\frac{1}{4}$-inch or $\frac{1}{2}$-inch, so there will be a considerable number of men whose recorded heights correspond to class-limits. It is therefore essential to show clearly in which class they are placed.

It should be noted that if each man's height is measured correct to, say, the nearest $\frac{1}{4}$-inch, an observation of 62 inches really means "$61\frac{3}{4}$ and under $62\frac{1}{4}$," $63\frac{1}{2}$ inches means "$63\frac{1}{4}$ and under $63\frac{3}{4}$," and so on. But the class "62 and under 64" includes recorded measurements of 62, $62\frac{1}{2}$, 63 and $63\frac{1}{2}$ inches only; consequently the correct description would be "$61\frac{3}{4}$ and under $63\frac{3}{4}$," i.e. the frequency table has overstated the class-limits by $\frac{1}{4}$-inch. On the other hand, Table 9.3 overstates the limits by £1 because it is the departmental practice to round off to the nearest £1 below. This may be important, particularly in dealing with averages (see Chapter 10).

Frequency Diagrams

There are various methods of representing frequency distributions graphically. In the case of a discrete variable with a small number of possible values, the simplest way is to erect vertical lines at regular intervals along the base line, with lengths proportional to the corresponding frequencies. Thus, for a graph of Table 9.1, points on the base line could be marked off at intervals of, say, $\frac{1}{4}$-inch or $\frac{1}{2}$-inch, representing 0, 1, 2 . . . 8 children, and vertical lines erected at these points to heights, of, say, 1·79, 2·05 . . . 0·01 inches, as in Fig. 9.1.

Alternatively columns of equal thickness could be erected instead of lines, on the principle of the bar chart. For a discontinuous variable this method is generally preferable, and to make it still clearer the actual frequency can be written over each column if desired.

Histograms

The above methods are unsuitable for a continuous variable, and it is best to regard the base line as an x-axis, on which *every* point represents a possible value of the variable. It is now impracticable to erect vertical lines or columns since the actual values of the items are generally not specified but only located in a range. In

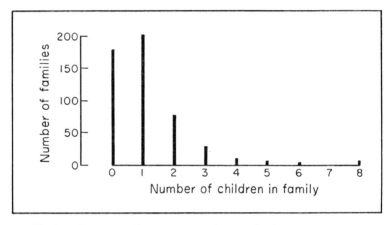

Fig. 9.1 Frequency diagram representing distribution of families by
number of children (see Table 9.1)

such a case the class-frequencies may be represented by means of a
histogram.

To construct this, points are marked off on the base line corresponding to the class-limits, so that the base line—or at least part
of it—is divided into a convenient number of sections corresponding to the class-intervals and proportional to them in length.
On each section is drawn a rectangle whose *area* represents the
corresponding class-frequency. The result is a series of rectangles
whose total area represents the total number of items observed.
These rectangles comprise the histogram, which forms a continuous area provided there are no missing classes, or classes with
frequency zero.

Fig. 9.2 shows a histogram of the height distribution given in
Table 9.2. As the reader will see later, the general form is one that
occurs frequently in statistics.

Notice that the two extreme classes, which have no definite
interval in the table, have been treated in the histogram as having
the same class-interval as the others, viz. 2 inches. As the numbers
in these two classes are very small, little accuracy has been lost in
this way. The only reasonable alternative was to omit them al-
together. When the class-intervals are not all equal the problem is
a little more difficult. The alternatives are either to omit the extreme
classes entirely (provided the numbers are small) or to assign to
them reasonable but arbitrary class-intervals.

Notice also that the ordinate in Fig. 9.2 is described as "Number
of men per class-interval." The description is only legitimate when
the class-intervals are equal, since it is the area, not the height, of

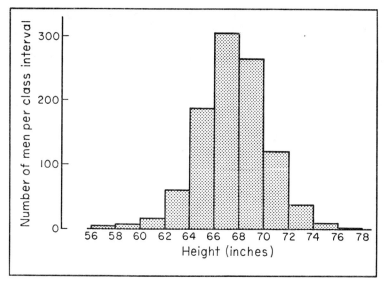

Fig. 9.2 Histogram representing frequency distribution of heights of 1,000 men (see Table 9.2)

the rectangle that represents the frequency. It comes to the same thing when the class-intervals are equal, but not otherwise, as the following simple illustration will demonstrate.

Fig. 9.3(a) shows part of a histogram with equal class-intervals, and Fig. 9.3(b) shows the effect of combining rectangles B and C into one. Obviously the height of the new rectangle should be intermediate between those of B and C, and this will in fact happen if the area

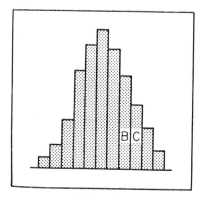

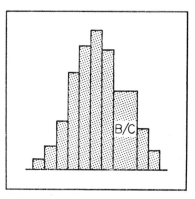

9.3(a) Histogram

Fig. 9.3(b) Histogram of Fig. 9.3 but with rectangles B and C combined is proportional to the frequency, so that the area of the new rectangle is the sum of areas B and C. But if the histogram had been constructed so that the *height* represented the class-frequency, the height of the new rectangle would have been the sum of the heights of B and C.

Frequency Polygons

When the class-intervals are all equal, the data can be represented by a frequency polygon. As in the histogram, the base line is divided into sections corresponding to the class-intervals, but instead of the rectangles, vertical lines are erected at the middle points of the sections with heights proportional to the class-frequencies. The points at the top of successive verticals are joined up, the extreme points being joined to the base line at the mid-points of what would have been the next sections. If squared paper is used, the points can be plotted in the usual way and there is no need for the vertical lines.

The frequency polygon is now rarely used. The histogram is preferable as a rule, since it shows more clearly the relation of each class-frequency to the total.

Frequency Curves

If the quantity of data is very large, and a small class-interval is chosen so that the number of classes is quite large, it often happens that the points of the frequency polygon follow a fairly regular course, so that a smooth curve could be drawn through, or at least near, all the points. The larger the number of classes and items the nearer, as a rule, will the polygon approach a smooth curve. This *frequency curve*, as it is called, may be regarded as the limiting form of the frequency polygon or of the histogram.

It may happen, of course, that the histogram or frequency polygon shows great irregularity, in which case there is no justification for attempting to draw a curve. On the other hand, many distributions appear to follow a fairly regular law, and in some cases there are good theoretical grounds for expecting them to do so. The data can then be smoothed by means of a frequency curve, with the confidence that if further observations were taken they would still conform closely to the curve. Examples are heights and weights of men and women, rates of mortality and sickness, errors of observation, and many kinds of biological data.

Common Types of Frequency Curve

Some of the commoner forms of frequency curve are described below. Most of this section applies equally to histograms and frequency polygons, but it will be convenient to discuss distributions in terms of frequency curves.

(a) *The Symmetrical "Hump-backed" Distribution*

This type of curve is of great importance in the theory of statistics. The distribution is symmetrical about a central maximum, the frequencies falling away to zero on either side (see Fig. 9.4).

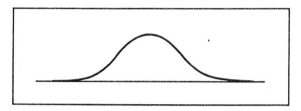

Fig. 9.4 Symmetrical hump-backed frequency curve

In practice it is unusual to find a perfectly symmetrical curve, but distributions of human heights, examination marks and experimental errors approximate to this form.

(b) *The Skew "Hump-backed" Distribution*

This is the commonest type of distribution occurring in statistical work. As before, the frequencies tail off to zero on either side of a peak, but they do so faster on one side than on the other, as in Fig. 9.5. Hence, it is called a *skew* or *asymmetrical* distribution. Good

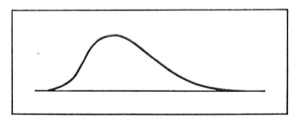

Fig. 9.5 Skew hump-backed frequency curve

examples of this type are the weight distribution of men or women, and the distribution of their ages at the time of marriage.

(c) *The "J-shaped" Distribution*

This type exhibits a maximum at one end of the range and tails away to zero at the other (see Fig. 9.6). It is often an extreme form of the second type, in which the interval between the peak and one end is so small that it has been impossible to analyse this part of the distribution so as to bring out the real form of the curve. Distributions of income and wealth follow this form.

These three types do not exhaust all possible forms of distribution, but they are well-recognized types and others may generally be regarded as variants of them.

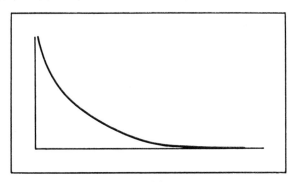

Fig. 9.6   J-shaped frequency curve

Cumulative Frequency Distributions

It is sometimes desirable to use a table of cumulative frequencies rather than that of class-frequencies. Suppose, for example, that it is required to estimate from Table 9.2 the number of men less than 5 ft 3 in. tall. The table might be rearranged as follows—

Table 9.5   Cumulative frequency distribution of 1,000 men by height
(Data of Table 9.2)

Height (inches)	Number of men	Cumulative number
Under 58	2	2
58 and under 60	5	7
60 and under 62	14	21
62 and under 64	60	81
64 and under 66	187	268
66 and under 68	304	572
68 and under 70	263	835
70 and under 72	121	956
72 and under 74	36	992
74 and under 76	7	999
76 and over	1	1,000
Total	1,000	—

The table shows that there are 21 men under 62 in. and another 60 of 62 in. or over but under 64 in. Assume that these 60 are spread evenly over this range; this may not be quite accurate, but it is the simplest assumption to make. This gives 30 men under

63 in. besides the 21 under 62 in., an estimated total of 51 men under 5 ft 3 in.

In practice this procedure would not be necessary to answer such a simple question, but it will serve as an illustration. Table 9.5 could, of course, have been compiled without the middle column, only the cumulative frequencies being shown and the height being given as under 58 in., under 60 in., under 62 in., and so on.

The summation may begin from the upper end of the range, as in the case of large incomes or estates.

These cumulative frequency tables may be represented by cumulative frequency polygons or curves, in which the height corresponds to the cumulative frequency. The cumulative frequency

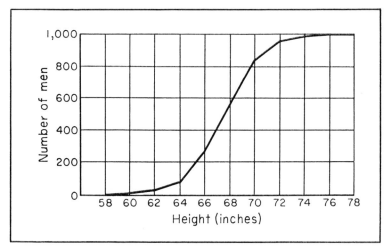

Fig. 9.7   Cumulative frequency polygon (from Table 9.5)

polygon derived from Table 9.5 is shown in Fig. 9.7. (Strictly speaking, it is not a polygon, but a series of short straight lines.)

A casual glance will show that it is much easier to draw a smooth curve through the points on this figure than in the case of the ordinary histogram or frequency polygon. This advantage is even more marked when the class-intervals are not all equal.

Such curves are known as *distribution curves* or *ogives* (see Fig. 9.8). Among other things, they can be used to estimate graphically the numbers of items with values less (or perhaps greater) than some specified value. Just as Table 9.5 was used to estimate the number of men less than 5 ft 3 in. in height, so a similar estimate could be obtained from the corresponding ogive by reading off the height corresponding to 63 in. In fact, this method would probably give a more accurate estimate than interpolation from the table.

If desired, the cumulative frequencies can be expressed as percentages of the total and the distribution curve drawn as a percentage cumulative frequency curve. Table 9.5 could easily be put into this form by inserting a decimal in each number in the last column, and similarly with Fig. 9.8, but the total is not usually such a convenient number. However, it is quite a simple task to calculate the percentages with the aid of a slide rule or a calculating machine.

One mistake frequently made by students is to plot the cumulative frequency against the mid-point, instead of the upper limit, of the last interval to be added. Thus, they would plot 268 men against

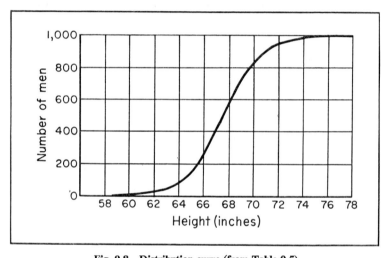

Fig. 9.8 Distribution curve (from Table 9.5)

65 in., confusing the procedure with that on the original distribution. With the cumulative distribution, however, this is wrong; 268 is the number of men under 66 in., not the number under 65 in. The shape of the curve is not affected; the effect of the mistake is simply to shift the whole curve half an interval (in this case one inch) to the left.

Exercises

9.1 Draw histograms (frequency bar charts) in such a way as to compare and contrast the distributions, according to size of vehicle, at end 1952 and end 1961. State any assumptions you make in drawing the histograms. Comment on the main changes between 1952 and 1961.

Number of road goods vehicles at end year

Unladen weight		Thousands of vehicles	
Over	Not over	1952	1961
	12 cwt	85	57
12 cwt	1 ton	277	506
1 ton	$1\frac{1}{2}$ tons	110	200
$1\frac{1}{2}$ tons	2 tons	106	102
2 tons	$2\frac{1}{2}$ tons	160	137
$2\frac{1}{2}$ tons	3 tons	142	191
3 tons	4 tons	39	121
4 tons	5 tons	16	46
5 tons	6 tons	7	17
6 tons	7 tons	3	9
7 tons	8 tons	2	10
8 tons		2	9

Source: Annual Abstract of Statistics

(IoT)

9.2 Illustrate the following data by means of a histogram, a frequency polygon and a frequency curve.

Monthly salary £	Number of employees
31–35	2
36–40	3
41–45	4
46–50	9
51–55	12
56–60	7
61–65	4
66–70	1

How does the frequency curve for this distribution differ from a curve showing skewness? (ACCA)

9.3 The following frequency distributions show the duration (reckoned in whole days) of absence from two causes of sickness in a certain transport organization—

Duration (days)	Number of absences	
	A	B
1	1	—
2	2	—
3	4	6
4	6	8
5	9	13
6	13	16
7	20	17
8	22	15
9	16	11
10	7	6
11	—	5
12	—	3

Draw the following three diagrams—

　　For "A"—Histogram and Frequency Polygon.
　　For "B"—Histogram and Frequency Curve (freehand).
　　For "A" and "B"—Ogive (two curves).　　　　　　　　(IoT)

9.4 Construct histograms to illustrate the following unemployment figures.

Duration of unemployment in weeks	Males	Females
One or less	46,675	13,202
Over 1 and up to 2	37,606	9,229
Over 2 and up to 3	23,671	4,935
Over 3 and up to 4	22,107	3,941
Over 4 and up to 5	20,422	3,774
Over 5 and up to 6	19,320	3,612
Over 6 and up to 7	18,542	3,440
Over 7 and up to 8	15,237	3,086
Over 8 and up to 9	14,555	2,727
Over 9 and up to 13	47,046	9,186
Over 13 and up to 26	80,893	13,344
Over 26 and up to 39	40,144	4,986
Over 39 and up to 52	25,424	3,002
Over 52	82,534	8,308
Total	494,176	86,772

Source: Employment and Productivity Gazette

(IoS)

9.5 A survey of people waiting at various bus stops included the following 70 timings (each measured to the nearest minute) of the number of minutes the people observed had to wait for a bus—

0	5	0	1	2	1	2
1	0	6	3	5	3	4
3	1	7	8	9	5	5
7	2	17	0	2	4	1
23	4	1	5	6	6	3
1	5	19	4	1	0	5
2	18	3	14	5	2	6
5	0	5	15	3	5	3
6	1	3	12	7	6	10
1	3	1	2	0	4	1

Choosing suitable, but not necessarily equal, class intervals, compile a grouped frequency distribution. By cumulating your frequency distribution estimate the missing figures in the following statement—

"... per cent of the people observed waited no more than one minute for a bus, but half of them had to wait ... minutes or more and ... per cent had to wait ten minutes or more."

In addition to the 70 timings recorded there were 3 untimed cases where people gave up and went away after waiting more than ten minutes. What would the missing figures in the statement become if these 3 people were allowed for in your estimates?   (IoT)

9.6 From the following table of income distribution, prepare a cumulative frequency curve (ogive) and explain the purpose of presenting the data in this form.

Range of incomes £	Number of families
1,150–1,249	5
1,250–1,349	4
1,350–1,449	8
1,450–1,549	13
1,550–1,649	13
1,650–1,749	11
1,750–1,849	11
1,850–1,949	9
1,950–2,049	5
2,050–2,149	6
2,150–2,249	3
2,250–2,349	1
2,350–2,449	1
Total	90

(ACCA)

10

Averages

If an examiner were asked how the boys' marks in a certain examination compared with those of the girls, he could reply in various ways. He might show the inquirer the actual marks obtained, but unless there happened to be a very striking difference between the sexes this method would be practically useless. He might produce the frequency distributions of the boys' marks and girls' marks separately, but even then a concise comparison might be difficult. Most likely, he would say something like this: "The average score of the boys was 3·4 more (or less) than that of the girls, but the girls' marks were rather more spread out."

The reader will see from this simple illustration that it is often desirable to go further than the compilation of a frequency distribution, and to condense the data into two or three quantities or *parameters* characterizing the distribution: first, a measure of central tendency, a typical value round which the various items are grouped, i.e. an *average*; secondly, a measure of *dispersion*, i.e. some indication of the way in which these items are scattered about the average; and possibly, although not very often, a measure of *skewness*, the tendency of items to spread out more on one side of the average than on the other. Other parameters are sometimes used in advanced work, but they are beyond the scope of this book.

This chapter will be confined to the averages; dispersion and skewness are left to Chapter 11.

The Arithmetic Mean

It is well known that the average of everyday speech is obtained by adding up, say, 47 items and dividing by 47. The word "average," however, is used in a wider sense in statistics, the particular average just mentioned being called the *Arithmetic Mean* (AM for short), or simply the *Mean*. This is the average generally used, but there are several others which, for various reasons to be discussed later, are sometimes preferable.

The formula for the mean is very simple. If $\bar{x}$ is the mean of n items $x_1, x_2, \ldots x_n$,

$$\bar{x} = \frac{1}{n}\{x_1 + x_2 + \ldots + x_n\}$$

or more briefly,

$$\bar{x} = \frac{1}{n}\Sigma x \qquad (10.1)$$

where the Greek letter Σ (capital sigma) denotes summation.

If $x_1, x_2, \ldots x_n$ occur with frequencies $f_1, f_2, \ldots f_n$ respectively, the formula becomes

$$\bar{x} = \frac{1}{N}\Sigma fx \qquad (10.2)$$

where $N(= \Sigma f)$ is the total frequency.

Calculating the Mean

If there are only a few items, it is easy to add them together and divide by their number. If there are many, however, it is better to form the frequency distribution, grouping them if necessary into a convenient number of classes and treating each class as if all its items had the same value, viz. the mid-value of the class-interval. The error involved in this method will be small unless there are large class-frequencies at the extremes.

Example 10.1

The scores of 145 competitors in a golf tournament ranged from 66 to 89 as shown in the table below. Find the mean score.

Score	Frequency	Score	Frequency	Score	Frequency
66	1	74	4	82	5
67	–	75	15	83	7
68	–	76	13	84	5
69	4	77	12	85	7
70	4	78	7	86	1
71	6	79	17	87	1
72	7	80	11	88	1
73	9	81	7	89	1
				Total	145

If the scores are grouped in class-intervals of 5, with central values 65, 70, etc., the distribution can be rewritten as follows—

Score	Central value (x)	Frequency (f)	Product (fx)
63–67	65	1	65
68–72	70	21	1,470
73–77	75	53	3,975
78–82	80	47	3,760
83–87	85	21	1,785
88–92	90	2	180
Total		145	11,235

$$\text{Mean} = \frac{11,235}{145} = 77 \cdot 5 \text{ approximately.}$$

Short-cut Methods

The student may wonder whether all this arithmetic is really necessary. The working can, in fact, be shortened considerably by (i) estimating the mean and taking an arbitrary origin somewhere near it, and (ii) taking the class-interval as the unit. Thus a score of 75 might be taken as an origin, or working mean, and differences from it measured in units of 5, giving the following table—

Central value (x)	Deviation in units (d)	Frequency (f)	Product (fd)	
65	− 2	1	− 2	
70	− 1	21	− 21	
75	—	53	—	—
80	+ 1	47		+ 47
85	+ 2	21		+ 42
90	+ 3	2		+ 6
Total		145	− 23 + 95 = + 72	

This gives +72/145 for the mean value of d, i.e. for the deviation of the mean score from 75, measured in class-intervals of 5. Hence the mean score is

$$75 + \left(\frac{72}{145} \times 5 \right) = 77 \cdot 5, \text{ as before.}$$

There are several pitfalls in these short-cut methods, and the student must be on his guard until practice has made perfect. He must watch his signs carefully, i.e. the signs of the deviations, the products fd and their algebraic sum. It is best to sum the negative and the positive products separately. Then he must remember to multiply the mean value of d by the class-interval before adding it to the working mean. He may wonder whether this method is really worth while or whether it is, like many other short cuts, "the longest way round." When he comes to the following chapter, however, he will realize that it is even more useful in the calculation of the standard deviation.

The Weighted Mean

One advantage of the arithmetic mean is that if the means of several sets of items and the number of items in each set are known, the mean of all sets combined can easily be calculated. For instance, suppose the mean age of 20 boys in a class is 14 years and that of 10 girls is 15 years. Then the total of the boys' ages is 280 years and that of the girls' ages 150 years, giving a total for the whole class of 430 years, so the mean age of the class is 430/30 years, or 14 years 4 months. And in general, if f_1 items have a mean m_1, f_2 a mean m_2, and so on, the mean of all the items is

$$\frac{f_1 m_1 + f_2 m_2 + \ldots}{f_1 + f_2 + \ldots} \quad \text{or} \quad \frac{\Sigma fm}{\Sigma f} \tag{10.3}$$

This leads to the *weighted mean*, as opposed to the simple or unweighted mean of a number of items. Suppose in the previous example the average age of the boys was known to be 14 years and that of the girls 15 years, but it was not known how many boys or how many girls there were. It would be necessary to assume that their numbers were nearly equal, to take the simple mean of 14 and 15 and to say that the average age of the class was (as nearly as could be judged) 14 years 6 months. But if it was known that there were about twice as many boys as girls, their averages could be "weighted" with factors 2 and 1 respectively, giving the average age of the class as

$$\frac{14 \times 2 + 15 \times 1}{2 + 1} = 14\tfrac{1}{3} \text{ years, i.e. 14 years 4 months.}$$

It is evident that if the boys outnumber the girls in the ratio 2 : 1 exactly, this method gives the correct answer. If the ratio is not exact but approximately correct the result will not be far out, as the student should verify for himself by taking, say, 19 boys and 11 girls, or 23 boys and 10 girls.

In general, the weighted mean of n items x_1, x_2 ... x_n, with weights w_1, w_2 ... w_n, is

$$\frac{w_1x_1 + w_2x_2 + \ldots + w_nx_n}{w_1 + w_2 + \ldots + w_n} = \frac{\Sigma wx}{\Sigma w} \qquad (10.4)$$

Compare this formula with (10.3). Evidently the mean obtained from a frequency distribution is a weighted mean in which the weights are the class-frequencies.

Weighted means are used chiefly in connexion with index numbers (see Chapter 16), but they are sometimes used to combine several observations or averages as in the following example.

Example 10.2

Three men, A, B and C, conduct an experiment to estimate a certain quantity and obtain values of 11·27, 11·41 and 11·36. On account of their differences in skill and experience, their observations are allotted weights of 5, 3 and 2 respectively (in everyday speech, A's estimate carries more weight than the others), and the weighted mean is taken as the best possible estimate of the item concerned.

Taking differences from 11·3, the weighted mean is

$$\frac{5(-0·03) + 3(0·11) + 2(0·06)}{5 + 3 + 2} = 0·03$$

Hence the quantity is estimated as 11·33.

The Median

Sometimes it is desirable to use a simpler average than the mean, one that is more easily calculated or for other reasons more suitable for the purpose in hand. One such average is the *Median*, defined as the central value of the variable. Thus if 37 items are arrayed in order of magnitude, the median is the value of the middle one, i.e. of the 19th. If there are 38 items, there are two that have an equal claim to be central, the 19th and 20th, so the median is taken as the average of these two. To take a simple example, if five men shoot and score 48, 55, 59, 67 and 70 the median score is 59, but if a sixth man scores 62, the median is $\frac{1}{2}$ (59 + 62), i.e. 60$\frac{1}{2}$.

When the data are grouped, matters are not quite so simple, as the median can only be estimated, not exactly located. The following example will show the method to be employed and should be studied carefully.

Example 10.3

Find the median height of the frequency distribution shown in Table 9.2.

The cumulative frequency table is first constructed as in Table 9.5 until half the total frequency has been reached, which in this case occurs in the sixth group. It is unnecessary to proceed further with the table unless the median is being found graphically (see below).

Height (inches)	Number of men	Cumulative number of men
Under 58	2	2
58 and under 60	5	7
60 and under 62	14	21
62 and under 64	60	81
64 and under 66	187	268
66 and under 68	304	572

Obviously the median is in the group 66–68 in., and common sense suggests that it is nearer 68 in. than 66 in. The best assumption to make is that the heights of the 304 men in this group are evenly spread over the interval. Now there are 500 men below the median (M, say) and 268 men below 66 in., so on the above assumption the distance (M–66) in. is (500–268)/304 of the class-interval which is 2 inches.

$$\text{Hence } M = 66 + \frac{500-268}{304} \times 2$$

$$= 67 \cdot 53 \text{ in. approximately.}$$

The assumption made here is not strictly correct, but the error involved is generally small.

It is quite easy to estimate the median graphically, and where items are not distributed evenly over the class containing it this method often gives a more accurate result. Construct the ogive and read off the value of the variable corresponding to the point whose height is half the total frequency. Fig. 10.1 shows the ogive for the data of Example 10.3, but continuing after the sixth group. The value of x corresponding to $y = 500$ is 67·5 in. as nearly as can be judged.

The Mode

A third very simple average is the *Mode*, the value of the variable occurring most often, corresponding to the highest point of the frequency curve. The mode is difficult to determine accurately from grouped data unless the number of data is very large. With a discrete variable it can usually be found at once, although if the frequencies are fairly small, peculiar results may occur through sampling errors (see Chapter 13). If the variable is continuous,

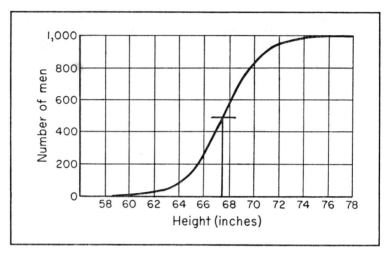

Fig. 10.1 **Estimation of the median of a grouped distribution (from Table 9.5)**

it is necessary to estimate what value would have the greatest frequency if there were enough data. In the case of Table 9.2, for instance, it is not at all clear whether, if each man's height were measured correct to 1/10 in., the mode would be 67·5 in., 67 in. or 68 in., or any other particular value. The best way to estimate the mode would be to fit a smooth curve to the data and take the variable corresponding to the peak.

Relation between the Mean, Median and Mode

When the frequency distribution is perfectly symmetrical, these three averages all coincide. If, however, the distribution is skew, they may differ appreciably. For most frequency curves of the "cocked-hat" type, the median lies between the mean and the mode, the distances being such that—

Mean − mode = 3 (mean − median) approximately.

In difficult cases, this relation may be used to estimate the mode of a grouped distribution of the "cocked-hat" type. It is not exact, but it works out remarkably well in practice. It is helpful to remember that when the longer tail of the distribution lies on the left, the mean, median and mode are in dictionary order and that the median lies nearer to the mean than to the mode—again as in the dictionary. When the longer tail is on the right the order is, of course, reversed.

Fig. 10.2 illustrates these three averages with reference to a frequency curve. The verticals M_1, M_2 and M_3 correspond respectively to the mean, the median and the mode. The latter gives the highest point of the curve, M_1 passes through the centre of gravity of the area, and M_2 divides the area into two equal parts.

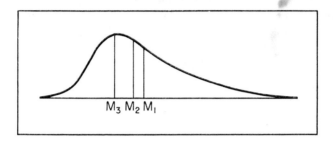

Fig. 10.2 Frequency curve showing mean, median and mode

Relative Advantages and Disadvantages

Before proceeding to consider other possible averages, it will be useful to examine the relative merits of the three already discussed.

The mean possesses several great advantages: it is easily understood and, as a rule, easily calculated, and it takes every item into account equally. The median is also easy to understand and to calculate; indeed it is often much easier to calculate than the mean, but it depends only on the middle items or groups, as does the mode, and is not affected by any change in the extreme items. Both the mean and the median are uniquely determined by the items, except for small errors in the case of grouped data, whereas the mode is generally difficult to estimate. There may indeed be more than one mode.

The chief advantage of the mean, however, is that it lends itself to algebraic treatment or to further calculations. It has been shown how, when two or more distributions are combined into one, their means can be weighted to find the mean of the whole. This cannot be done with the median or the mode.

The mean also has the advantage that the total value of the population can be worked back from the number of items and their mean value. For instance, if the average wage of 1,000 men is £28 a week, the weekly wage bill is £28,000. This could not be deduced from knowledge of the mode or median. There are, of course, cases where the total is of little interest; if the mean age of a class of 30 boys is 10 years, it is not very useful to know that their total is 300 years. But if their mean pocket money is 35p a week, it is useful to know that their total pocket money is £10·50, particularly to the master responsible for distributing it to them.

It must not be supposed, however, that all the advantages are with the mean. If the population contains a few exceptional cases, they may distort the mean unduly; for example, it would be misleading to quote the mean income of a small community if it happened to include a millionaire. Again, the median may be chosen for economic reasons; if a batch of 25 lamps is being tested by the time taken to burn out, it will save both time and material if the time taken for the 13th lamp to burn out is taken as the average life of a lamp. Again, where the extreme groups of a distribution have no definite range, it is difficult to determine the mean accurately. The median is unaffected by this.

These three averages having been discussed at some length, it only remains to consider briefly two of minor importance.

The Geometric Mean

The *Geometric Mean* (GM) of n observations $x_1, x_2, \ldots x_n$ is given by the formula

$$g = \sqrt[n]{(x_1 x_2 \ldots x_n)} \tag{10.5}$$

It is generally calculated by taking logarithms, since

$$\log g = \frac{1}{n} \Sigma \log x \tag{10.6}$$

Example 10.4

Find the geometric mean of the numbers 104, 119, 86, 94, 155 and 77.

Number	Log
104	2·0170
119	2·0755
86	1·9345
94	1·9731
155	2·1903
77	1·8865
	6)12·0769
GM = 103·0	2·0128

The student should find the AM and compare it with the GM.

The GM is used almost entirely in the calculation of index numbers (see Chapter 16), and should only be used when all the items are positive. It has the effect of reducing the influence of large items and increasing that of small ones. It is easily proved (the proof is found in many algebra books) that the GM is always less than the AM.

The Harmonic Mean

The *Harmonic Mean* (HM) of n observations $x_1, x_2, \ldots x_n$ is given by the formula

$$\frac{1}{h} = \frac{1}{n} \sum \frac{1}{x} \tag{10.7}$$

Thus, as with the GM, finding the HM reduces to finding the AM with a simple transformation of the variable. For the GM we take logarithms, for the HM we take reciprocals.

Example 10.5

Find the harmonic mean of 20 and 30,

$$\frac{1}{h} = \frac{1}{2}\left(\frac{1}{20} + \frac{1}{30}\right) = \frac{5}{120} = \frac{1}{24}$$

$$\therefore \quad h = 24$$

The usual illustration is average speed over a return journey. Thus, a cyclist might cover a distance d miles at an average speed of 20 mph and the same distance back at an average speed of 30 mph. The total time taken is $d/20 + d/30$ hours. It must also be $2d/h$ hours, where h mph is the average speed over the double journey. Example 10.5 shows that $h = 24$, the HM of 20 and 30.

In practice, the HM is very seldom used. It is only mentioned here because examiners sometimes set questions on it.

Exercises

10.1 The following figures represent the weekly expenditure on fruit and vegetables by each of 60 families in a sample.

£	£	£	£	£	£
0·98	0·71	1·05	0·79	1·53	1·38
1·24	1·76	0·78	0·89	0·91	1·03
0·78	0·84	0·91	0·93	1·12	1·25
1·26	0·87	0·98	0·74	0·51	0·38
0·33	1·19	1·48	1·62	0·41	0·48
0·69	0·88	1·08	1·03	1·78	0·69
0·51	0·58	1·28	1·51	1·33	0·66
0·74	0·51	1·58	1·17	1·79	0·64
1·61	0·48	1·16	1·26	1·44	1·83
0·46	0·36	0·21	1·96	2·13	0·87

You are required to

(*a*) arrange the figures in the form of a grouped frequency distribution;

(*b*) calculate the average weekly family expenditure on fruit and vegetables, using the short-cut method;

(*c*) plot the figures in the form of a histogram.

(based on ICWA)

10.2 Calculate the mean strength of the following frequency distribution of lots of cotton yarn:

Strength of lots of cotton yarn	
Pounds per square inch	Number of lots
70–74	1
75–79	6
80–84	17
85–89	29
90–94	20
95–99	17
100–104	13
105–109	10
110–114	6
115–119	3
120–124	2
125–129	1

State what your answer indicates and compare the mean with the median as a statistical average. (ACCA)

10.3 The following figures give the daily flows through a sewage plant, in thousands of gallons, on 35 consecutive days. Arrange them in a frequency distribution and calculate *three* measures of central tendency.

118, 110, 152, 168, 160, 168, 119, 92, 175, 152, 168, 144, 144, 184, 152, 168, 168, 104, 137, 144, 170, 124, 114, 202, 187, 192, 221, 192, 168, 175, 175, 181, 211, 168, 211. (IoS)

10.4 The arrival time of a certain train is recorded on 25 separate occasions. It is late on each occasion, the number of minutes it is late being given in Table A opposite. After steps are taken to improve its timekeeping the time lost on the next 25 occasions is as in Table B.

Calculate the arithmetic mean and the median for each group of observations. Do your results enable you to judge whether the train's timekeeping has improved? Find an alternative way, by further calculation or graphically, of illustrating the differences between the two groups of figures.

Table A			Table B		
2	33	13	5	25	3
1	18	28	14	17	23
8	13	31	7	2	15
6	8	2	10	9	21
23	1	10	22	20	17
14	4	27	17	9	6
38	1	12	18	5	8
29	22		3	23	
18	32		19	17	

(IoT)

10.5 Calculate the median of the following frequency distribution and also estimate the percentage of all the cars travelling at 33 miles per hour or more.

Miles per hour	Number of cars (f)
15·0–19·9	4
20·0–24·9	16
25·0–29·9	40
30·0–34·9	18
35·0–39·9	6
40·0 and over	4

What does the median value of this distribution indicate, and what are the advantages of the median as a statistical average? (ACCA)

10.6

Mangolds: estimated yield per acre

County	Tons	County	Tons
Bedfordshire	32·7	Derbyshire	20·9
Berkshire	24·4	Devon	27·1
Buckinghamshire	31·5	Dorset	34·5
Cheshire	18·5	Durham	20·9
Cornwall	26·0	Essex	24·2
Cumberland	19·0	Kent	25·6

Source: Agricultural Statistics 1963/4

Calculate the mean, median and mode of the above data. Which average do you here consider to be the most useful and why?

(IoS)

10.7 Define the weighted arithmetic mean of a set of observations x_1, x_2, . . . x_n with weights W_1, W_2, W_3 . . . W_n.

The following data refer to a group of observations—

Commodity	Weight	Observation
a	4	113
b	5	140
c	3	120
d	W_d	130
e	W_e	108
f	W_f	105

Given that the weighted arithmetic mean of the six observations is 117·68 and that the sum of the six weights is 25, calculate the weighted arithmetic mean of commodities (i) a, b, and c, (ii) d, e, and f.

(IoS)

10.8 Define in your own words the arithmetic mean, the geometric mean. and the harmonic mean. Calculate each of these measures using the following figures—

3, 8, 12, 6, 5, 17, 9, 15

Give an example to show when you would use the harmonic mean in preference to the arithmetic mean, and explain why you would do so. (IoT)

11

Dispersion and Skewness

AN AVERAGE is not in itself an adequate summary of a frequency distribution. It is also necessary to have some idea of the way individual items are scattered about that average. Average income, for example, does not mean very much if great wealth and poverty exist side by side in a community. Also, as will appear later (Chapter 13), it is often necessary to measure the variability of a distribution in order to estimate the accuracy of a test result or to assess its significance.

The Range

A rough and ready measure of dispersion is the difference between the highest and lowest values observed, called the *range*. It is not very satisfactory because it depends solely on the two extreme values and may be very misleading if there are one or two abnormal items, e.g. if one of a group of examination candidates is a genius. Moreover, since exceptional items are more likely to occur when the number of items is large, the range tends to increase as the number of items increases. In some cases, e.g. where there are large open groups at the ends like "65 years and over," it is impossible to estimate the range. There are, however, circumstances in which the average range of small groups of items provides a good estimate of the dispersion (see Chapter 15).

Values of Position

A good way of showing how items are distributed is to find the values that divide the distribution into a number of equal parts when the data are arranged in order of magnitude. For example, the nine *deciles* divide the distribution into ten equal parts, one-tenth of the items falling below the first decile, two-tenths below the second, and so on.

Percentiles go further and divide the distribution into 100 equal parts when the data are very numerous. They are greatly favoured by psychologists, and the "percentile rank" is often used when the percentile itself, as a numerical value, does not exist. Thus boys could be assessed in respect of capacity for leadership, say, and a percentile rank of 85 would mean that a boy excelled 85 per cent of his fellows in that respect.

Quartiles

The most commonly used values of position are the *quartiles*, which divide the data into four equal parts. The second quartile has already been introduced as the median, by which name it is always known. The other two are called the lower and upper quartiles and are generally denoted by Q_1 and Q_3 respectively.

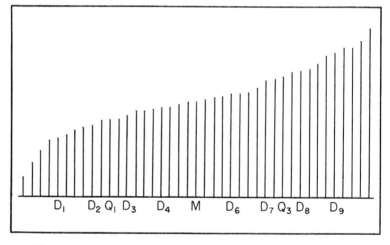

Fig. 11.1   Array of forty-one items showing positions of quartiles and deciles

Fig. 11.1 shows 41 items represented graphically in the form of an array, i.e. arranged in order of magnitude. It also shows the deciles and quartiles. The lower quartile is the value of the 11th item, since there are 10 below and 30 above; similarly the upper quartile is the value of the 31st item. The deciles have been taken as the values of the 5th, 9th . . . 37th items. The median is shown as *M*.

Had there been only 40 items, the lower quartile could have been taken as the average of the 10th and 11th, as there would then be 10 items below it and 30 above; similarly with the upper quartile. When there are 42 or 43 items, the position is not so clear, as no value will divide the distribution exactly in the ratio 1 : 3. Probably the best rule in such cases is to take the quartiles as corresponding to the positions $\frac{n+1}{4}$ and $\frac{3(n+1)}{4}$, where n is the number of items, rounding these quantities to the nearest integer where necessary. Thus, if $n = 42$, the quartiles will be the values of the 11th and 32nd items; if $n = 43$, the values of the 11th and 33rd items.

Statisticians are not unanimous about this, but if there are many items it makes no appreciable difference whether one value or the next or the average of the two is taken as the quartile; if there are few items, the error involved is small compared with errors of sampling (see Chapter 13).

Quartiles of a grouped distribution

When the data are grouped, the quartiles cannot normally be determined accurately; they can only be estimated as the median is estimated (see page 95), either by interpolation or by means of the ogive.

Example 11.1

To find the quartiles of the height distribution given in Table 9.2.

The method is the same as in Example 10.3, where the object was to find the median. The cumulative frequency table, shown completed this time as in Table 9.1, is as follows—

Height (inches)	Number of men	Cumulative number of men
Under 58	2	2
58 and under 60	5	7
60 and under 62	14	21
62 and under 64	60	81
64 and under 66	187	268
66 and under 68	304	572
68 and under 70	263	835
70 and under 72	121	956
72 and under 74	36	992
74 and under 76	7	999
76 and over	1	1,000

Evidently Q_1 is in the range 64 to 66 in. and as there are 250 items below Q_1,

$$Q_1 = 64 + \frac{250-81}{187} \times 2 = 65.81 \text{ in.}$$

Similarly,

$$Q_3 = 68 + \frac{750-572}{263} \times 2 = 69.35 \text{ in.}$$

The interpolation method shown above assumes that the items in any particular class-interval are evenly distributed. No such assumption is made in the graphical method. The ogive for the above table is shown in Fig. 11.2.

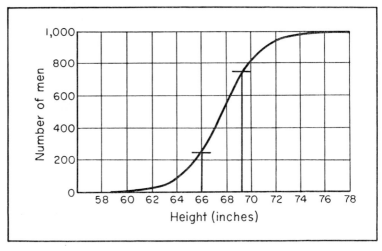

Fig. 11.2 Distribution curve showing the estimation of the quartiles (from Table 9.5)

As nearly as can be judged from the graph,

$$Q_1 = 65.8 \text{ in.}, \ Q_3 = 69.2 \text{ in.}$$

The median has already been estimated (Example 10.3) as 67·5 in., so these results could be expressed in the following form—

25% of the men are under 65·8 in.
50%   „   67·5 in.
75%   „   69·2 in.

or alternatively,

25% of the men are under 65·8 in.
25% of the men are between 65·8 and 67·5 in.
25% of the men are between 67·5 and 69·2 in.

and

25% of the men are over 69·2 in.

Quartiles, deciles and similar quantities—quantiles, as they are sometimes called—are commonly used in connexion with incomes, wages, and social inquiries generally.

The Quartile Deviation

The quantity $\frac{1}{2}(Q_3-Q_1)$ is called the *quartile deviation*, or less often, the *semi-interquartile range*. There is no recognized symbol for it, but it is commonly abbreviated to QD. When the distribution is symmetrical, the QD is equal to the difference between the median and either quartile.

Using the data of Example 11.1,

$$QD = \tfrac{1}{2}(69\text{·}2\text{--}65\text{·}8) \text{ in.} = 1\text{·}7 \text{ in.}$$

The Mean Deviation

The *Mean Deviation* differs from the range and the QD in that it takes every item into account. It is the arithmetic mean of all deviations from the mean (or alternatively from the median), counting all such deviations as positive. This proviso is necessary because the algebraic sum of all deviations from the mean is zero.

Consider the numbers 19, 22, 36, 27, 16. Their mean is 24, and their deviations from the mean are −5, −2, +12, +3 and −8, which add up to zero, as they must. Ignoring the signs, the MD is $\frac{1}{5}(5 + 2 + 12 + 3 + 8) = 6\text{·}0$. In practice, such a small number of items would not call for a measure of dispersion, but the example is given for simplicity.

At first sight this seems an ideal method of measuring dispersion. What could be more natural, having found the mean, than to take the average difference from it? Unfortunately, the MD has certain serious disadvantages and is therefore seldom used. In particular, it does not lend itself to further calculation or algebraic treatment as the mean does for averages. Difficulties also arise in computing the MD of a grouped frequency distribution.

Sometimes the MD is measured from the median. It is easily seen that the mean deviation is minimum when measured from this point, for if the origin of measurement is moved from the median to some other point, some deviations will be reduced, but at least as many will be increased each by the same amount.

The Standard Deviation

None of the three possible measures of dispersion described above (the range, the QD and the MD) lends itself to mathematical treatment. Here the *Standard Deviation* has the advantage over them, as will be seen. That is the chief justification for what at first sight seems a clumsy and troublesome quantity to compute. Instead of the deviations themselves being averaged, as in the MD, the deviations from the mean are squared, then the sum of the squares is divided by the number of items, and the square root of the quotient is the standard deviation.

The SD is denoted by the Greek letter σ (small sigma, not to be confused with the summation sign Σ, which is capital sigma), and σ^2 is called the *variance* of the distribution. If there are n items $x_1, x_2, \ldots x_n$, with a mean $\bar{x}$, then by definition

$$\sigma^2 = \frac{1}{n}\{(x_1 - \bar{x})^2 + (x_2 - \bar{x})^2 + \ldots + (x_n - \bar{x})^2\}$$

$$= \frac{1}{n}\Sigma(x - \bar{x})^2 \qquad (11.1)$$

To take the trivial example of the previous section, the SD of the numbers 19, 22, 36, 27 and 16 is given by

$$\sigma^2 = \tfrac{1}{5}\{(-5)^2 + (-2)^2 + 12^2 + 3^2 + (-8)^2\}$$
$$= \tfrac{1}{5}(25 + 4 + 144 + 9 + 64)$$
$$= 49 \cdot 2$$
$$\therefore \sigma = \sqrt{49 \cdot 2} = 7 \cdot 01 \text{ approximately.}$$

Normally the SD would not be calculated for only five items. It has been done here to explain the process as clearly and simply as possible, because it is essential that this very important measure and the method of calculating it should be thoroughly understood.

In the above example the mean was a whole number and the sum of squares was easily calculated. Generally, however, the mean is a fraction or decimal, and to simplify the arithmetic deviations are measured from the origin or some value near the mean, and an adjustment made to correct this. If the working mean is taken as origin, the formula is

$$\sigma^2 = \frac{1}{n}\Sigma x^2 - \bar{x}^2$$

The proof of this formula is as follows
By (11.1) above

$$n\sigma^2 = \Sigma(x - \bar{x})^2 = (x_1 - \bar{x})^2 + (x_2 - \bar{x})^2 + \ldots + (x_n - \bar{x})^2$$

Expanding the contents of the brackets and noting that $\bar{x}$ is a constant,

$$n\sigma^2 = \Sigma x^2 - 2(\Sigma x)\bar{x} + n\bar{x}^2$$

But

$$\Sigma x = n\bar{x}$$

$$\therefore n\sigma^2 = \Sigma x^2 - 2n\bar{x}^2 + n\bar{x}^2$$

$$= \Sigma x^2 - n\bar{x}^2$$

and

$$\sigma^2 = \frac{1}{n}\Sigma x^2 - \bar{x}^2 \qquad (11.2)$$

Substituting $\frac{1}{n}\Sigma x$ for $\bar{x}$ gives the alternative form

$$\sigma^2 = \frac{1}{n}\left\{\Sigma x^2 - \frac{1}{n}(\Sigma x)^2\right\} \qquad (11.3)$$

where

$$(\Sigma x)^2 = (x_1 + x_2 + \ldots + x_n)^2$$

This form is usually more convenient for numerical work, for, even with large numbers, Σx and Σx^2 are easily computed with a calculating machine. Incidentally, formula (11.2) shows that the sum of squares Σx^2 is least when the deviations are measured from the mean.

If x_1 occurs f_1 times, x_2 f_2 times, and so on, the formulae become

$$\sigma^2 = \frac{1}{N}\Sigma f x^2 - \bar{x}^2 \qquad (11.2a)$$

and

$$\sigma^2 = \frac{1}{N}\left\{\Sigma f x^2 - \frac{1}{N}(\Sigma f x)^2\right\} \qquad (11.3a)$$

where $N(= \Sigma f)$ is the total number of items.

Example 11.2

To find the standard deviation of the following distribution of motor drivers having 0, 1, 2, etc. accidents over a period of years.

No of accidents	No of drivers
0	45
1	36
2	40
3	19
4	12
5	8
6	3
7	2
8	1
Total	166

There is no point here in taking a trial mean, or even in calculating the mean. The calculation is as follows

x	f	fx	fx²
0	45	—	—
1	36	36	36
2	40	80	160
3	19	57	171
4	12	48	192
5	8	40	200
6	3	18	108
7	2	14	98
8	1	8	64
Total	166	301	1,029

The "crude sum of squares"

$$= \Sigma fx^2 = 1,029$$

and the "correction term"

$$= \frac{1}{N} (\Sigma fx)^2 = \frac{301^2}{166} = 545 \cdot 8$$

∴ "Corrected sum of squares"

$$= 1,029 - 545 \cdot 8 = 483 \cdot 2$$

$$\therefore \sigma^2 = \frac{483 \cdot 2}{166} = 2 \cdot 91$$

$$\therefore \sigma = 1 \cdot 71 \text{ approximately.}$$

The Standard Deviation of a Grouped Distribution

With a grouped distribution it is generally best to take the mid-point of one group as a working mean and measure deviations from it with the class-interval as unit, or, if the class-intervals are not all equal, with some other convenient unit.

Example 11.3

To find the SD of the distribution of golf scores given in Example 10.1, grouped in intervals of 5 (see pages 91 and 92).

Central value	x	f	fx	fx²
65	− 2	1	− 2	4
70	− 1	21	− 21	21
75	—	53	—	—
80	+ 1	47	+ 47	47
85	+ 2	21	+ 42	84
90	+ 3	2	+ 6	18
Total		145	+ 72	174

Crude sum of squares

$$= 174$$

Correction term

$$= \frac{72^2}{145} = 35 \cdot 75$$

Corrected sum of squares

$$= 138 \cdot 25$$

$$\therefore \sigma = \sqrt{\frac{138 \cdot 25}{145}} = 0 \cdot 976 \text{ units}$$

Since deviations have been measured in units of 5 strokes,

$$SD = 0 \cdot 976 \times 5 = 4 \cdot 88 \text{ strokes}$$

The true SD, calculated from the original data, is 4·54 strokes. Considering that the grouping is so broad and that all but three items are in four groups only, this difference is to be expected. It can, however, be reduced by what is called "Sheppard's correction for grouping." This consists in deducting $\frac{1}{12}h^2$ from the mean square before taking the square root, h being the class-interval. In the above example, taking the class-interval as unit,

$$\sigma^2 = \frac{138 \cdot 25}{145} - \frac{1}{12} = 0 \cdot 870$$

$$\therefore \sigma = 0 \cdot 933 \text{ units}$$
$$= 4 \cdot 66 \text{ strokes}$$

This correction applies only to distributions of the hump-backed type and not, for example, to J-shaped distributions. The justification for the value $1/12$ is beyond the scope of this book.

Comparison of the Various Measures of Dispersion

Apart from the range, which for reasons already stated is an unreliable measure of dispersion, the easiest measure to compute and to interpret is the QD. Unfortunately it does not lend itself easily to mathematical treatment, but it is sometimes preferable to the MD or SD for a grouped distribution when the extreme groups are large or have no definite range. The MD is also easy to understand and easy to calculate for a moderate number of items, but grouped data may be troublesome.

Only the SD is suitable for algebraic manipulation. For instance, two or more distributions with known means and standard deviations can be combined into one distribution whose mean and SD can be calculated. The SD can also be used in precise mathematical tests of significance (see Chapter 13), and it can be proved that for the hump-backed type of distribution it is the most efficient measure of dispersion, i.e. it can be estimated from a sample more accurately than any other. It is difficult to calculate, however, when a grouped distribution has unequal class-intervals or extreme groups with no definite range or mean.

There is no definite relation between the QD, MD and SD, but in a hump-backed distribution that is either symmetrical or only moderately skew, the QD is about $\frac{2}{3}\sigma$ and the MD about $\frac{4}{5}\sigma$. It is evident that the same measure must be used to compare several distributions, e.g. it is no use comparing the QD of one with the SD of another.

Coefficient of Variation, etc.

It is sometimes less important to measure the absolute dispersion than to measure the relative dispersion, i.e. to measure the dispersion as a ratio or percentage of the average instead of in units of length, weight, etc. A standard deviation of one ounce would matter far less in weighing out 7 lb boxes than it would for $\frac{1}{4}$ lb packets of tea.

The commonest measure of relative dispersion is the *Coefficient of Variation*, v, obtained by dividing the SD by the mean and expressing it as a percentage. In short,

$$v = 100 \frac{\sigma}{\bar{x}} \tag{11.4}$$

The coefficient of variation is not dependent on choice of units. Indeed, it can be used to compare the dispersion of distributions expressed in different units, e.g. the daily output of two factories producing different commodities, one expressed in tons and the other in yards.

Other similar measures are MD/mean or MD/median and $(Q_3 - Q_1)/(Q_3 + Q_1)$. These are rarely, if ever, used in practice.

Example 11.4

Brown has a batting average of 32 with a SD of 13. Green has an average of 47 with a SD of 18. Which is the more consistent player?

It is easily shown that $13/32 > 18/47$. Hence Green has the smaller coefficient of variation and may be considered more consistent than Brown.

The coefficient of variation and similar measures cannot be used if the origin is arbitrary. Thus it would be ridiculous to use the coefficient of variation to compare movements of temperature in towns with different mean temperatures. Indeed, for any one place, the result would differ according to the scale used, e.g. Centigrade, Fahrenheit or absolute, because $0°$ is not the same temperature in the three cases. In fact, this coefficient only makes sense if all items are positive magnitudes.

Skewness

Even when the average and dispersion are known it is possible to fit many distributions of different shapes to them. One question that sometimes arises is: "Is the distribution symmetrical or lopsided, and if the latter, how lopsided is it?" The answer is generally given in the form of a *coefficient of skewness*, an abstract quantity independent of the unit in which the variable is measured.

Fig. 11.3 (which is a repetition of Fig. 10.2) shows a typical skew distribution, with a longer "tail" on the right of the mode than on the left. In this case the skewness is said to be positive.

When the longer tail is on the left the skewness is negative. When skewness is positive the mean (M_1), median (M_2) and mode (M_3) are in that order, starting from the right, as in Fig. 11.3.

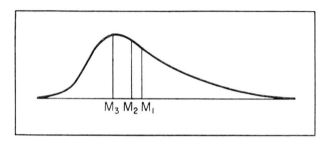

$$M_3\ M_2\ M_1$$

**Fig. 11.3   Positive skew distribution showing positions of Mean (M_1)
Median (M_2) and Mode (M_3)**

The more lopsided the distribution, the more will the mean, median and mode differ, so one possible coefficient of skewness is given by

$$j_1 = \frac{M_1 - M_3}{\sigma}, \tag{11.5}$$

or, since $M_1 - M_3 = 3(M_1-M_2)$ approximately,

$$j_1 = \frac{3(M_1 - M_2)}{\sigma} \tag{11.6}$$

An alternative measure is based on the fact that in a skew distribution the median is not exactly midway between the quartiles. The coefficient is given by

$$j_2 = \frac{Q_1 + Q_3 - 2M_2}{Q_3 - Q_1} \tag{11.7}$$

The latter has the advantage that it must lie between -1 and $+1$, whereas j_1 has no such convenient limits. There are more theoretically satisfactory ways of measuring skewness, but they have no place in elementary work.

Examples on skewness are often based on J-shaped distributions, which are so obviously skew that there is no point in measuring the skewness. In any case, formulae (11.5) and (11.6) do not apply to such distributions, but principally to the moderately skew "cocked-hat" type. In the author's opinion, however, measures of skewness are mainly of academic interest.

Exercises

11.1

Numbers of buses and coaches by seating capacity			
Number of seats		31 December 1962	31 December 1966
Over	Not over		
	14	2,360	2,768
14	32	4,748	2,724
32	40	12,584	7,213
40	48	16,788	20,022
48	56	20,956	15,407
56	64	13,553	14,190
64		5,303	12,442
Total		76,292	74,766

Source: Passenger Transport in Great Britain

Draw two cumulative frequency curves in a single chart and estimate from them the median seating capacity of buses and coaches in 1962 and in 1966 and the quartile deviation in each year. Say what the chart and your calculations tell you about the changes between the two years. State what assumption you made about the final class interval and what effect this assumption had upon your results. (IoT)

11.2 The following table indicates the marks obtained by 100 students in an examination—

Marks	Frequency
55– 64	1
65– 74	2
75– 84	9
85– 94	22
95–104	33
105–114	22
115–124	8
125–134	2
135–144	1

Using linear interpolation, calculate the median mark obtained. Describe the type of distribution for which the median is the best measure of location, and indicate whether the above distribution is of this type. Also calculate the upper and lower quartiles.   (IoS)

11.3 Plot the following figures on a cumulative frequency graph and from this find the value of all the deciles of the distribution.

Distribution of a sample of women by stature

Stature	Number in sample
under 4' 9"	39
4' 9" and 4' 10"	273
4' 11" and 5' 0"	798
5' 1" and 5' 2"	1,176
5' 3" and 5' 4"	1,206
5' 5" and 5' 6"	584
5' 7" and 5' 8"	217
5' 9" and above	56
Total	4,349

Source: *Women's Measurements and Sizes*
(H.M.S.O. 1957)

How will the values of the deciles help a clothing manufacturer decide on a sizing system?   (ICWA)

11.4 Calculate the mean deviation of the following frequency distribution.

Weight of fowl (lb)	Number of fowls
0·0– 4·9	3
5·0– 9·9	9
10·0–14·9	15
15·0–19·9	30
20·0–24·9	18
25·0–29·9	12
30·0–34·9	9

Indicate what your answer conveys about the weight of birds in this distribution.   (ACCA)

11.5 The following table gives the unladen weight of road goods vehicles in 1967.

Unladen weight (tons)	Vehicles (thousands)
0–1	545
1–2	406
2–3	169
3–5	274
5–8	99
Over 8	34
Total	1,527

Calculate the mean unladen weight and the standard deviation. Comment on your results. (IoT)

11.6 (a) Calculate the standard deviation of the weekly wages earned by a group of employees of Stitchers Limited, from the following data—

Wages £	Number of employees
1– 3	1
4– 6	4
7– 9	9
10–12	6
13–15	2
16–18	3

(b) What does a knowledge of the standard deviation of this group of workers' wages add to our information about their earnings? (ACCA)

11.7 A sample of 35 values has mean 81·52 and standard deviation 3·74. A second sample of 85 values from the same population has mean 82·31 and SD 3·62. Find the mean and SD of the combined sample of 120 values. (IoS)

11.8 The following figures give sets of marks for 10 candidates in two examinations, the English paper being marked out of 20 and the

Arithmetic paper out of 50. Using an appropriate measure, find whether there is relatively more variation in the marks for one examination than in those for the other.

Candidate	English marks	Arithmetic marks
1	17	45
2	8	30
3	4	9
4	20	25
5	11	17
6	19	33
7	5	41
8	16	24
9	10	37
10	7	15
Total	117	276

(IoS)

12

Probability and Some Special Distributions

PROBABILITY is of fundamental importance in the theory of statistics, but it can only be treated here in a very elementary manner. The reader who is interested in the subject will find it dealt with more thoroughly in textbooks on algebra or in more advanced books on statistical theory.

Books or chapters on probability deal very largely with problems in coin-tossing, dice-throwing and games of chance, not because it is a trivial or frivolous subject but because the principles of probability are most easily explained by means of such problems. One great advantage they have is that the chances of, say, heads and tails can be assumed known, whereas in most practical problems the probabilities can only be estimated. Thus, a man's expectation of life can be calculated only from current mortality rates, which may not apply to future generations, or from assumed rates in which some arbitrary allowance is made for future improvements.

Definitions and Fundamental Laws

In discussing events which can have either of two possible results, it will be convenient to refer to these results as success and failure. The probabilities of success and failure are usually denoted by p and q respectively.

If an event can result in n different ways, all equally likely, of which l are successes and m are failures,

$$p = \frac{l}{n} \text{ and } q = \frac{m}{n}$$

Since

$$l + m = n,$$
$$p + q = 1 \tag{12.1}$$

For example, if a playing-card is drawn at random from a full pack, the chance of drawing a spade is $\frac{13}{52}$ ($= \frac{1}{4}$), and the chance of drawing a card other than a spade is $\frac{39}{52}$ ($= \frac{3}{4}$).

More generally, if an event can have one of n possible results and their respective probabilities are $p_1, p_2, \ldots p_n$,

$$p_1 + p_2 + \ldots + p_n = 1 \tag{12.2}$$

It may be taken as self-evident that if p_1 is the probability of result A, and p_2 that of result B, where A and B are *mutually exclusive* (i.e. cannot happen together), then the probability of "either A or B" is $p_1 + p_2$. Similarly for three or more results. This is the *addition law* of probability.

Thus, if a number is being chosen at random from the numbers 1 to 100, the chance of picking an even number is $\frac{1}{2}$, that of picking a number ending in 5 is $\frac{1}{10}$, and that of picking either an even number or a number ending in 5 is $\frac{1}{2} + \frac{1}{10} = \frac{3}{5}$. It would, however, be untrue to say that the chance of picking either an even number or a multiple of 5 was $\frac{1}{2} + \frac{1}{5}$, because both events include numbers ending in 0, i.e. they are not mutually exclusive. In plain English, they overlap.

The other basic law is the *multiplication law* of probability, which states that if $p_1, p_2, p_3 \ldots p_n$ are the probabilities of success in n independent events, the probability of success in all together is $p_1 p_2 p_3 \ldots p_n$. Thus, if the probability that a man aged 60 will survive ten years is $\frac{3}{4}$ and that of a woman aged 50 surviving ten years is $\frac{8}{9}$, the probability that they will both survive ten years is $\frac{3}{4} \times \frac{8}{9} = \frac{2}{3}$.

It is assumed here that the events are independent, i.e. that the result of one does not affect the result of another. Successive tosses of a coin are independent, for however many times it may come down "heads," the probability of the next result being "heads" is still $\frac{1}{2}$ (unless it is biased in some way); and if A's neighbour wins a football pool, A's chance of winning it the following week is neither increased nor diminished. But if cards are drawn from a pack and not replaced, the chance of drawing a red card at any particular stage will depend on the colours of the cards already drawn.

Conditional Probability

The multiplication law is easily adapted to dependent events. This is best shown by an example.

Example 12.1

A class contains 21 boys and 15 girls, assumed to be equal in attainments. Find the probability that a boy will take first prize and a girl second prize.

The probability that a boy will take first prize is $\frac{21}{36}$. The probability then that, out of the other 35 children, one of the 15 girls will take second prize, is $\frac{15}{35}$. The probability of the double event is therefore

$$\frac{21}{36} \times \frac{15}{35} = \frac{1}{4}$$

An alternative argument is as follows: a boy can be chosen for first prize in 21 ways, and a girl for second prize in 15 ways, giving a total of 21 × 15 ways. Disregarding sex, the two prizes can be awarded in 36 × 35 ways. The required probability is therefore

$$\frac{21 \times 15}{36 \times 35} = \frac{1}{4} \text{ as before.}$$

To solve the great majority of problems in probability, the reader needs only the above basic theorems, a logical mind, and plenty of practice in working out examples.

The Probability of *r* Successes in *n* Independent and Similar Events

The present section is of great importance in the theory of sampling. Suppose an event is repeated and that the probability p of success is the same each time. The problem is to find the respective probabilities of 0, 1, 2, . . . n successes in n trials.

To take a concrete case, suppose a six-sided die is thrown repeatedly, and that a six counts as a success, any other value being a failure, so that $p = \frac{1}{6}$ and $q = \frac{5}{6}$.

Consider first the case of two throws—or two dice thrown together, which comes to the same thing. By the multiplication law, the probability of two successes is $(\frac{1}{6})^2$ and that of two failures is $(\frac{5}{6})^2$. The probability that the first throw is a success and the second a failure is $\frac{1}{6} \times \frac{5}{6}$. The probability that the first throw is a failure and the second a success is also $\frac{1}{6} \times \frac{5}{6}$. Hence the probability of one success and one failure, regardless of order, is $2 \times \frac{1}{6} \times \frac{5}{6}$.

It is easily verified that

$$\left(\frac{1}{6}\right)^2 + \left(\frac{5}{6}\right)^2 + 2\left(\frac{1}{6}\right)\left(\frac{5}{6}\right) = 1$$

On similar reasoning, the chances of 0, 1, 2, 3 successes in three throws are respectively

$$\left(\frac{5}{6}\right)^3, \quad 3\left(\frac{5}{6}\right)^2\left(\frac{1}{6}\right), \quad 3\left(\frac{5}{6}\right)\left(\frac{1}{6}\right)^2, \quad \left(\frac{1}{6}\right)^3$$

In general, if there are n trials, and p and q are the chances of success and failure in each trial, the probability of r successes and $(n-r)$ failures *in a specified order* is $p^r q^{n-r}$. But since r successes can occur in ^nC_r ways, the probability of r successes regardless of their order is $^nC_r p^r q^{n-r}$, where

$$^nC_r = \frac{n!}{r!(n-r)!}$$

Those familiar with the Binomial Theorem will recognize the probabilities of 0, 1, 2, ... n successes as the respective terms in the expansion of $(q + p)^n$, viz.

$$q^n, npq^{n-1}, \ldots {}^nC_r p^r q^{n-r}, \ldots p^n \qquad (12.3)$$

Example 12.2

Find the probability that out of 10 coins tossed, exactly 3 will be "heads."

Since $p = q = \frac{1}{2}$, the required probability is

$$^{10}C_3(\tfrac{1}{2})^3(\tfrac{1}{2})^7 = \frac{10.9.8}{1.2.3}(\tfrac{1}{2})^{10}$$

$$= \frac{15}{128}$$

Probability Distributions

The expressions given by formula (12.3) provide an example of a *probability distribution*, which is a series of probabilities corresponding to all possible values of a variable, in this case the number of successes in n events. Probability and frequency distributions are closely connected, for if a large number of experiments are performed, the relative frequencies of the various possible results will approximate to the probabilities. A probability distribution may therefore be regarded as a frequency distribution standardized so that the total frequency is unity, and it can be represented by a histogram, polygon or curve like any other frequency distribution.

Fig. 12.1 shows, in the form of a frequency polygon, the probability distribution of scores of 3, 4, ... 18 with three dice thrown simultaneously. The actual probabilities are

$$\frac{1}{216}(1, 3, 6, 10, 15, 21, 25, 27, 27, 25, 21, 15, 10, 6, 3, 1)$$

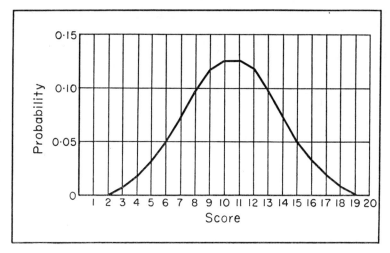

Fig. 12.1 Probability distribution of total scores made by throwing three dice

The fraction outside the bracket means that all numbers inside the bracket should be divided by 216. The calculation of these probabilities is rather laborious.

When the variable is continuous, it is still possible to compile and plot a probability distribution, but the interpretation is not quite so simple. It is, however, possible to estimate the probability that the height or weight of a man chosen at random lies between certain limits, or that a machine will break down within a stated time.

Fig. 12.2 shows a possible distribution for times run by a machine before it breaks down. Since the total of all probabilities must be unity, the curve is scaled so that the area between it and the axes is also unity. The area $PNMQ$ then represents the probability that the machine will break down between $1\frac{1}{2}$ and 2 hours, the times represented by N and M. The area to the left of PN gives the probability of its breaking down within $1\frac{1}{2}$ hours, and the "tail" to the right of QM that of the machine running at least 2 hours without a breakdown.

Unfortunately these areas cannot be read off as vertical heights can, and before a continuous probability distribution can be of much use it must be tabulated. Several such distributions, like the so-called Normal Distribution (page 126), have been extensively tabulated, as also have some discrete distributions, so that the probability of the variable taking certain values or lying within or outside a given range can be found with reasonable accuracy.

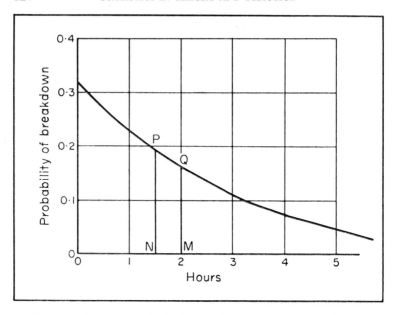

Fig. 12.2 Probability distribution of times taken for a machine to break down

Expectation

The mean of a probability distribution is often called the *expected value*, or *expectation*, of the variable x, and is denoted by $E(x)$. If x can take values x_1, x_2, ... x_n with respective probabilities p_1, p_2, ... p_n, then—

$$E(x) = p_1x_1 + p_2x_2 + \ldots + p_nx_n \qquad (12.4)$$

It is easily seen that $E(x)$ will be the average value of x in a very large number of observations.

"Expected value" is rather a misnomer, since it may be a value the variable cannot possibly take. For example, if five coins are tossed, the "expected number" of heads is $2\frac{1}{2}$. The term "expectation" is preferable for this reason.

Example 12.3

A boy answers four questions "yes" or "no" and guesses each time, not knowing the answers. He is to receive 5p for 2 correct answers, 15p for 3, and 30p for 4. Find his expectation.

From formula (12.3), since $p = q = \frac{1}{2}$,

probability of 2 correct answers $= {}^{4}C_{2}(\frac{1}{2})^{4} = \frac{3}{8}$

,, ,, 3 ,, ,, $= {}^{4}C_{3}(\frac{1}{2})^{4} = \frac{1}{4}$

,, ,, 4 ,, ,, $= {}^{4}C_{4}(\frac{1}{2})^{4} = \frac{1}{16}$

$\therefore$ expectation in pence

$= \frac{3}{8} \times 5 + \frac{1}{4} \times 15 + \frac{1}{16} \times 30$

$= \frac{15}{8} + \frac{15}{4} + \frac{15}{8}$

$= 7\frac{1}{2}p$

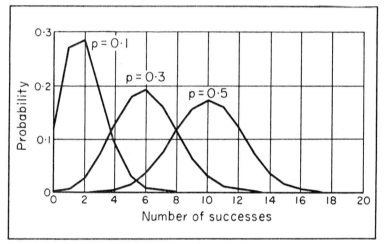

Fig. 12.3 Probability distribution of 0, 1, 2, . . . 20 successes in twenty
trials when $p = 0.1$, 0.3 or 0.5

The Binomial Distribution

One of the commonest probability distributions is the one already
given in formula (12.3) by the binomial expansion of $(q + p)^{n}$
and called the *Binomial Distribution*. This will now be considered
at some length on account of its importance.

When the chances of success and failure are equal, i.e. when
$p = q = \frac{1}{2}$, the probability of x successes is ${}^{n}C_{x}(\frac{1}{2})^{n}$, and equal to that
of x failures, or $(n-x)$ successes. Obviously the distribution is
symmetrical about $x = \frac{1}{2}n$.

When p and q are unequal, the distribution is skew, but if n is
fairly large and neither p nor q is very small, the skewness is only
slight and the tails can almost be ignored because the extreme
frequencies are so small. If p is smaller than q, the longer tail is on
the right.

Fig. 12.3 shows distributions corresponding to the expansion of

Table 12.1 Terms of the binomial expansion of 1,000 $(q + p)^{20}$
(Figures to nearest unit)

Number of successes	Frequency		
	$p = 0\cdot1$	$p = 0\cdot3$	$p = 0\cdot5$
0	122	1	—
1	270	7	—
2	285	28	—
3	190	72	1
4	90	130	5
5	32	179	15
6	9	192	37
7	2	164	74
8	—	114	120
9	—	65	160
10	—	31	176
11	—	12	160
12	—	4	120
13	—	1	74
14	—	—	37
15	—	—	15
16	—	—	5
17	—	—	1
18	—	—	—
19	—	—	—
20	—	—	—

$(q + p)^{20}$ when $p = 0\cdot1$, $0\cdot3$ and $0\cdot5$. The first $(p = 0\cdot1)$ is markedly skew, the second almost symmetrical, and the last perfectly symmetrical. The actual probabilities, multiplied by 1,000 to avoid decimals, are given in Table 12.1.

The mean of the binomial distribution is np, as is evident from the very definition of p. It will be shown later (page 136) that the standard deviation is given by the formula

$$\sigma^2 = npq \qquad (12.5)$$

This formula is extremely important, as the following chapter will show.

The Normal Distribution

It can be shown that when n is large and neither p nor q is very small, the binomial distribution approximates to a continuous distribution of supreme importance in the theory of statistics, called the *Normal Distribution*, or sometimes the *Gaussian Distribution*. The curve representing it is called the *Normal Curve* (see Fig. 12.4), and its equation, for the benefit of the mathematically minded students, is

$$y = \frac{1}{\sigma\sqrt{(2\pi)}} e^{\frac{-x^2}{2\sigma^2}} \tag{12.6}$$

It is symmetrical about the mean and stretches away to infinity on either side.

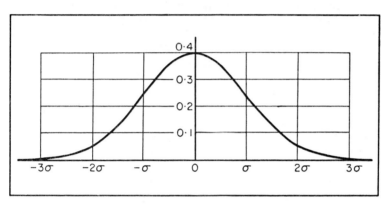

Fig. 12.4   Normal curve

As equation (12.6) shows, the Normal Distribution has the great advantage of containing only one arbitrary constant σ (the SD), which means that it is easy to tabulate. This can be done in various ways. Table 12.2 shows the proportion P of the area of the curve that lies to the right of the ordinate at various distances from the mean.

Table 12.2   Proportion of the area of the Normal Curve to the right of the ordinate at deviation x from the mean

x/σ	P	x/σ	P	x/σ	P
0	0·5000	1·0	0·1587	2·0	0·0228
0·1	0·4602	1·1	0·1357	2·1	0·0179
0·2	0·4207	1·2	0·1151	2·2	0·0139
0·3	0·3821	1·3	0·0968	2·3	0·0107
0·4	0·3446	1·4	0·0808	2·4	0·0082
0·5	0·3085	1·5	0·0668	2·5	0·0062
0·6	0·2743	1·6	0·0548	2·6	0·0047
0·7	0·2420	1·7	0·0446	2·7	0·0035
0·8	0·2119	1·8	0·0359	2·8	0·0026
0·9	0·1841	1·9	0·0287	2·9	0·0019
				3·0	0·00135

Clearly, if P is the area to the right of (mean $+ x$), $1 - P$ is the area to the left of it and $1 - 2P$ the area between the limits (mean $\pm x$). Thus, the area between (mean -1.5σ) and (mean $+1.5\sigma$) is

$$1 - 2(0.0668) = 0.8664.$$

Since these areas represent probabilities, the matter can be expressed in this way: the probability that an item picked at random from a Normal Distribution will exceed the mean by 1.5σ or more is 0.0668: the probability that it will differ from the mean by 1.5σ or more either way is $0.0668 \times 2 = 0.1336$.

The quartiles are at distances 0.6745σ from the mean. This quantity is sometimes called the *probable error*, a complete misnomer since it is not the most probable deviation and need not be an error. It is simply the deviation such that greater and smaller deviations are equally numerous. The term "probable error" is obsolete in statistics, but not yet in astronomy.

The Normal Curve was studied by Gauss in connexion with the theory of errors of observation, and is sometimes called the Curve of Errors. It is a kind of ideal curve which very nearly fits many of the distributions met in statistics, such as distributions of men's heights, examination marks and errors of observation. The student should, however, beware of statements sometimes made that the Normal Curve is found in "natural data." Very often a Normal Curve can be fitted to data, and unless they are very numerous it may not be possible to prove that the distribution is not Normal, but perfect Normality rarely occurs in practice.

The Normal Curve also has important mathematical properties, most of which are beyond the scope of this book. For example, a powerful technique called Analysis of Variance, much used in biological and industrial experiments, is based on the assumption that the data are Normal. Fortunately this and other methods can often be used with confidence even when there is a marked difference from the Normal Distribution.

The Poisson Distribution

Another limiting form of the binomial distribution is the *Poisson Distribution*. This is the form to which the binomial distribution approximates when n is large, p very small, and np of the order of unity, say less than 10. It can be shown (the proof will not be given here) that successive terms of the binomial distribution approximate to

$$e^{-m}, \; me^{-m}, \; \tfrac{1}{2}m^2e^{-m}, \; \ldots \; \frac{m^x}{x!} \, e^{-m}, \text{ etc.,}$$

where $m = np$ and e denotes the base of the system of natural logarithms (= 2·71828 . . .).

Since the mean and standard deviation of the binomial distribution are np and $\sqrt{(npq)}$ respectively, and q is now nearly equal to 1, it follows that the mean and standard deviation of the Poisson Distribution are respectively m and $\sqrt{m}$.

This distribution, like the Normal Distribution, has the advantage that it involves only one parameter, and can therefore be easily tabulated. It is useful when the expected number of successes or failures in a large number of events is small. For example, if the average number of defective articles in fairly large batches of a given size is 2, the probability of finding no defectives in a single batch is e^{-2}, the probability of finding exactly one is $2e^{-2}$, and so on. The actual probabilities are as follows

Number of defectives	Probability
0	0·1353
1	0·2707
2	0·2707
3	0·1805
4	0·0902
5	0·0361
6	0·0120
7	0·0034
8	0·0009
9	0·0002

The Poisson Distribution may arise in any situation where an event is liable to occur at random at irregular intervals of time or space, and the probability of occurrence in a small interval of given length or size is constant. Examples are—

(a) The number of accidents occurring in a given factory during a day or week.

(b) The number of cars passing a certain point in one minute.

(c) The number of telephone calls received on one line during a given short period.

(d) The number of organisms on a square of a haemocytometer.

In all these examples it is assumed that conditions, such as weather, which have any bearing on the event being observed are constant over the period or interval in question.

Exercises

12.1 (a) A pack of cards is cut at random, one card exposed, the suit of the card noted and the whole pack re-made. If this procedure is performed four times, what is the probability that none of the exposed cards is a spade?

(b) In an office, four letters and the four corresponding envelopes are typed. If the letters are put into the envelopes at random, what is the chance that none of the letters is put into the correct envelope?

(IoS)

12.2 The mean daily output of operatives in a spinning factory is Normally distributed with a mean of 540 yards and a standard deviation of 17 yards. The management are introducing more modern machines but will only train workers whose mean daily output is already 525 yards or more to use them. What percentage of the factory force can expect re-training? (IoS)

12.3 (a) Fit a Normal Curve to the following data—

Weight (nearest 1 lb)	Number of men
Under 100	6
100–119	43
120–139	93
140–159	191
160–179	263
180–199	212
200–219	144
220–239	40
Over 239	8
Total	1,000

(b) Estimate the weight which is exceeded by 10 per cent of the men.

(IoS)

12.4 It is specified that the bursting strength of a particular type of paper should be at least 25 pounds per square inch (psi). Current production is such that the mean bursting strength is at 26 psi, with a standard deviation of 0·75 psi.

(a) Assuming that bursting strength is distributed Normally, what proportion of production is outside specification?

(b) What percentage increase in bursting strength should be required to ensure that only 1 per cent of production is outside specification? (The standard deviation may be assumed to be unchanged.) (IoS)

12.5 (a) Under what conditions can the Binomial distribution be approximated by the Poisson distribution, and in what circumstances is such an approximation useful?

(b) If the proportion of defectives in a bulk is 4 per cent, find the probability of not more than one defective in a simple of 10.

(c) A shopkeeper's sales of a certain article amount to 4 per month on the average, the sales occurring at random and independently. He replenishes his stock once a month. To what number should he make up his stock in order to reduce the chance of running out of stock to less than 1/100? (IoS)

12.6 What is the Poisson distribution?

Fifty slug traps are set up in a regular array in a field of turnips, and one night 90 slugs are trapped. What is the expected number of empty traps if the Poisson distribution is fitted?

If 21 of the traps are empty, is this consistent with expectation? How would you explain the discrepancy, if any? (IoS)

12.7 A rolling machine in a paper mill produces on average one flaw every 500 feet of paper. Assuming that the number of flaws in a given length of paper has a Poisson distribution, find the probability that

(a) a 2,000 ft roll has no flaws,

(b) a 500 ft roll has at most three flaws,

(c) out of three 1,000 ft rolls, one has one flaw and the others have none. (IoS)

13

Sampling

IN Chapter 4 the technique of sampling was described in general terms, with special reference to sample inquiries of the social survey type. It was stated that the larger the sample and the more carefully it was chosen, the more reliable it would be as a cross-section of the population, but the numerical expression of the reliability of samples was deferred to this chapter.

The theory of sampling is generally divided into the sampling of attributes and the sampling of variables (see page 7). In the sampling of attributes, the first problem is to estimate the proportions of items falling into categories that cannot be expressed numerically, like brown-eyed, blue-eyed, etc., male and female, or alive and dead. In the sampling of variables the problem is to estimate the mean, and perhaps the dispersion, of some variable like height or weight or the yield of a product. In both cases subsequent problems are to assess the precision of these estimates and to judge the significance of differences between one estimate and another, or between an estimate and some expected value.

Throughout this chapter it will be assumed that all samples are random samples unless otherwise stated.

Standard Deviation of a Sum, Difference, or Mean

Suppose there are two variables x and y, each with its own distribution and with means m_x and m_y and standard deviations σ_x and σ_y respectively. Suppose further that an item is taken at random from each distribution and that the two items are added together, giving $z = x + y$. Then z will form another distribution with mean $m_z = m_x + m_y$. The problem now is to find σ_z, the standard deviation of z.

To simplify the calculations, let x, y and z be measured from their respective means, and suppose x and y can each take a limited number of values x_1, x_2, . . . y_1, y_2, etc., with probabilities p_1, p_2, . . . q_1, q_2, etc., respectively. (Students familiar with integral calculus can generalize this section to the case where the distributions of x and y are continuous. The non-mathematician may skip the proof and continue from formula (13.2).)

Then if $z_{rs} = x_r + y_s$, the probability associated with z_{rs} is $p_r q_s$. Since x, y and z are measured from their means,

$$\Sigma_r p_r x_r = \Sigma_s q_s y_s = \Sigma_r \Sigma_s p_r q_s z_{rs} = 0,$$
$$\sigma_x^2 = \Sigma_r p_r x_r^2 \text{ and } \sigma_y^2 = \Sigma_s q_s y_s^2 \tag{13.1}$$

Then

$$\begin{aligned}
\sigma_z^2 &= \Sigma_r \Sigma_s p_r q_s z_{rs}^2 \\
&= \Sigma_r \Sigma_s p_r q_s (x_r + y_s)^2 \\
&= \Sigma_r \Sigma_s p_r q_s x_r^2 + 2\Sigma_r \Sigma_s p_r q_s x_r y_s + \Sigma_r \Sigma_s p_r q_s y_s^2 \\
&= (\Sigma_r p_r x_r^2)(\Sigma_s q_s) + 2(\Sigma_r p_r x_r)(\Sigma_s q_s y_s) + (\Sigma_r p_r)(\Sigma_s q_s y_s^2) \\
&= (\sigma_x^2 \times 1) + (2 \times 0 \times 0) + (\sigma_y^2 \times 1) \quad \text{(from (13.1)}) \\
&= \sigma_x^2 + \sigma_y^2 \tag{13.2}
\end{aligned}$$

In other words, the variance of z ($= x + y$) is equal to the sum of the variances of x and y, provided that x and y are independent.

This extremely important formula can be extended to any number of distributions, irrespective of their form or the numbers of items they contain. Furthermore, it applies equally to differences, as can be seen by changing the sign of y in formula (13.1). Thus if

$$d = x - y,$$
$$\sigma_d^2 = \sigma_x^2 + \sigma_y^2 \text{ as before,} \tag{13.2a}$$

and if

$$X = x_1 \pm x_2 \pm \ldots \pm x_n,$$
$$\sigma_X^2 = \sigma_1^2 + \sigma_2^2 + \ldots + \sigma_n^2 \tag{13.3}$$

Two very important results can be obtained from equation (13.3) when x_1, x_2, . . . x_n are all taken *from the same distribution*. Since the standard deviation is now the same for each item ($= \sigma$, say),

$$\sigma_X^2 = n\sigma^2$$

or

$$\sigma_X = \sigma \sqrt{n} \tag{13.4}$$

Now $\bar{x}$ is the mean of the n items selected, $\bar{x} = X/n$, and $\sigma = \sigma_X/n$.

Hence

$$\sigma_{\bar{x}} = \sigma \frac{\sqrt{n}}{n}$$

$$= \frac{\sigma}{\sqrt{n}} \qquad (13.5)$$

To see what this means, suppose the heights of a large body of men, such as an army regiment, are distributed with a mean of 69·2 in. and a standard deviation of 2·7 in. Then if a large number of samples of 25 men are taken, the average heights of these samples will themselves form a distribution with a mean of 69·2 in. (the population mean) and a standard deviation not of 2·7 in. but of

$$\frac{2\cdot7}{\sqrt{25}} = \frac{2\cdot7}{5} = 0\cdot54 \text{ in.}$$

If samples of 100 men were taken, the standard deviation would be reduced to 0·27 in. Clearly formula (13.5) is a useful guide to the precision of an estimated average.

Sampling Distributions

Other quantities besides sums, differences and means can be estimated from samples. Sometimes variability is more important than the average, and the sample is then taken to obtain an estimate of the standard deviation; or some measure of skewness may be required to test whether the distribution is nearly Normal. Any such quantity estimated from a sample is called a *statistic*.

For a given size of sample any statistic will have a *sampling distribution* formed by all possible values of the statistic and depending on the number of items in each sample and the form of the parent distribution. For example, if the parent distribution is Normal, it can be shown that the distribution of the mean of n items is also Normal, but the standard deviation depends on the value of n (see Fig. 13.1). It can also be shown, but only in a more advanced book than this, that for most types of distribution the mean of n items becomes more and more Normally distributed as n increases.

Note that the parent distribution need not consist of items actually observed. It may be a probability distribution or a purely hypothetical distribution of values that might be observed if observations were continued indefinitely. For example, in an experiment to determine a physical constant, a temperature or the yield of a process under given conditions, the various results obtained by different observers, by different methods and with different apparatus may be regarded as a sample of an infinite population of possible estimates. Such a hypothetical population is often called a *universe*.

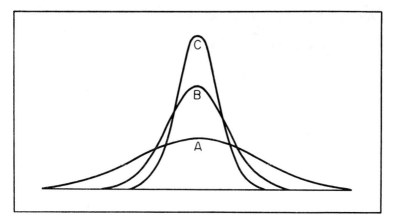

Fig. 13.1 Normal distribution *(A)*, with distributions of sample means of four items *(B)*, and of nine items *(C)*

The standard deviation of a sampling distribution is generally called the *Standard Error*. Thus in the example given in page 134, the standard error of the mean height of 25 men is 0·54 in. The word "error" is used in a technical sense to denote the difference between an estimate and the true value, which in this case is the average height of the population, and the standard error is a measure of the uncertainty of that estimate.

Standard Error of a Proportion

It is now possible to find the standard error of an estimated proportion, e.g. the proportion of births that result in twins. Consider a large population of N items, divisible into two categories only, which it is convenient to call successes and failures. (The reader can decide for himself into which category he would put twins.) Then if p and q are the (unknown) proportions of successes and failures respectively, there will be Np successes and Nq failures.

Suppose a sample of n items is taken to estimate p and q, n being reasonably large but small compared with N. If there are x successes in the sample, x/n will be taken as the estimate of p. The expected value of x is np (see page 126). The task now is to find its standard error.

This can be done very neatly by a device which, although at first sight it looks like sharp practice, is perfectly legitimate. Let the variable be the number of successes *in a single item*. The resulting frequency distribution is unusual, since the variable can take only two possible values, viz. 0, with frequency Nq, and 1, with frequency

Np (see Fig. 13.2). It is nevertheless a frequency distribution, and its standard deviation σ can be calculated in the usual way. The mean of the distribution is clearly p.

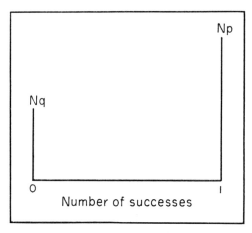

Fig. 13.2 Distribution of successes and failures in a population of size N

Then

$$N\sigma^2 = Nq(0 - p)^2 + Np(1 - p)^2$$
$$= Nqp^2 + Npq^2$$
$$= Npq(p + q)$$
$$= Npq$$
$$\therefore \sigma^2 = pq$$

But the sample is a sample of n items from this distribution, so from formula (13.4) the standard error of x is given by

$$\sigma_x^2 = npq$$
$$\sigma_x = \sqrt{npq} \qquad (13.6)$$

Since the observed proportion is x/n,
standard error of proportion

$$= \frac{1}{n} \sqrt{npq}$$
$$= \sqrt{\frac{pq}{n}} \qquad (13.7)$$

Confidence Limits

It is generally possible, using the standard error of a statistic such as a mean or proportion, to assess the probability that the true value, of which that statistic is an estimate, lies between certain limits, called *confidence limits*, or alternatively to find the limits corresponding to a specified probability. As already remarked, the mean of a large sample of observations is almost Normally distributed for most types of parent distribution; so also is the observed proportion of successes, provided neither np nor nq is very small, say less than 10. Consequently the tables of the Normal Distribution may be used with considerable accuracy to calculate the required limits. The following two examples will illustrate the method.

Example 13.1

It is desired to estimate the proportion of people in a certain town who have been vaccinated. Out of a random sample of 300, only 75 had been vaccinated. Find the 95 per cent confidence limits for the population.

Substituting $n = 300$, $p = 75/300 = 0.25$, $q = 0.75$ in formula (13.7),

$$\sigma_p = \text{SE of } p = \sqrt{\frac{0.25 \times 0.75}{300}} = 0.025.$$

From the tables of the Normal Distribution, 95 per cent of the observations lie within the range (Mean $\pm 1.96\sigma$). Hence with 95 per cent confidence, the proportion of the population that has been vaccinated lies between the limits

$$0.25 \pm 1.96 \times 0.025 = 0.25 \pm 0.049$$

i.e. between 0.201 and 0.299, or between 20.1 per cent and 29.9 per cent.

One difficulty that was glossed over in this example is that since it was the *estimated* values of p and q that were substituted in the formula, σ_p itself may not be correct. It is, however, the best estimate that can be found from the sample, and if the true value of p were 30 per cent, σ_p would only be increased to 0.0265. For greater precision a larger sample would of course be necessary.

If it were desired, following this preliminary survey, to estimate with 95 per cent confidence the percentage vaccinated correct to, say, 2 per cent, the size of sample required could be calculated as follows

$1.96\sigma_p$ must be made equal to 0.02

$$\therefore (1.96)^2 \frac{pq}{n} = (0.02)^2$$

Putting $p = 0.25$, $q = 0.75$ (the best estimates yet available)

$$n = \frac{(1.96)^2 \times 0.25 \times 0.75}{0.0004} = 1{,}800 \text{ approximately.}$$

The next sample is based on an illustration given earlier in this chapter.

Example 13.2

It is required to estimate the average height of a large body of men. A random sample of 400 men is found to have a mean height of 69·2 in. and a standard deviation of 2·7 in. Find the 99 per cent confidence limits of the true average height.

The standard error of the mean height is $2.7/\sqrt{400} = 0.135$ in. From the tables, 99 per cent of the Normal Distribution lies within 2.576σ of the mean, so with 99 per cent confidence the true mean lies in the *confidence interval*

$$69.2 \pm 0.135 \times 2.576 = 69.2 \pm 0.35, \quad \text{or} \quad 68.85 \text{ in. to } 69.55 \text{ in.}$$

As in the previous example, the standard error is subject to sampling errors because the standard deviation could be estimated only from the sample. With small samples the uncertainty of the standard deviation means that the tables of the Normal Distribution cannot safely be used. The next two sections show how the standard deviation is estimated from a small sample and how the confidence limits have to be modified.

Estimating the Standard Deviation

Let the sample consist of n observations $x_1, x_2, \ldots x_n$, where n is fairly small, say 25 or less. Let $\bar{x}$ be the sample mean, μ the population mean, supposed unknown, and σ the population standard deviation, also unknown. If μ were known, σ^2 could be estimated from $1/n \, \Sigma(x - \mu)^2$. If, instead, it were estimated from $1/n \, \Sigma(x - \bar{x})^2$ the estimate would be too low, for $\Sigma(x - \bar{x})^2$ is less than $\Sigma (x - \mu)^2$. In fact, by an extension of formula (11.2),

$$\Sigma(x - \bar{x})^2 = \Sigma(x - \mu)^2 - n(\mu - \bar{x})^2$$

Now the expected value of each $(x - \mu)^2$ is σ^2, and since $\sigma_{\bar{x}}^2 = \dfrac{\sigma^2}{n}$, the expected value of $n \, (\mu - \bar{x})^2$ is σ^2.

$\therefore$ Expected value of $\Sigma(x - \bar{x})^2 = n\sigma^2 - \sigma^2$

$$= (n - 1)\sigma^2$$

The best estimate of σ^2 is therefore $\Sigma(x - \bar{x})^2/(n - 1)$, and this statistic is generally denoted by s^2.

$$\therefore s = \sqrt{\frac{\Sigma(x - \bar{x})^2}{n - 1}} \tag{13.8}$$

The reader should note that some writers use s^2 to denote $1/n \Sigma(x - \bar{x})^2$, in which case the best estimate of σ^2 is $ns^2/(n - 1)$.

To recapitulate: if the "true," or population, values of the mean and standard deviation of a distribution are unknown, they can be estimated from a sample. The sample mean $(\bar{x})$ gives an unbiased estimate of the true mean μ, and the sample variance s^2 gives an unbiased estimate of the true variance σ^2, provided $n - 1$ is used as divisor for the latter—but not for the mean—instead of n.

"Student's t"

The standard error of the mean of a small sample can now be taken as $s/\sqrt{n}$, where s is given by formula (13.8). Unfortunately the factors by which this is multiplied to calculate confidence limits are also affected by the fact that σ is not precisely known. For example, at the 95 per cent confidence level, the appropriate factor is no longer 1·96 but something a little greater, depending on the value of n. This new statistic is denoted by t and is generally known as Student's t after W. S. Gosset, who first described its distribution, writing under the *nom de plume* "Student."

The mathematical theory of the t-distribution is beyond the scope of this book, but a few values of t are shown in Table 13.1 for

Table 13.1   Selected values of t

Degrees of freedom $(= n - 1)$	t	
	$P = 0·05$	$P = 0·01$
1	12·71	63·66
2	4·30	9·92
3	3·18	5·84
5	2·57	4·03
10	2·23	3·17
20	2·09	2·85
60	2·00	2·66

95 per cent and 99 per cent confidence levels. In the table, P denotes the probability that a value of t greater than the figure shown will occur by chance. The left-hand column represents the number of *degrees of freedom* or independent comparisons between observations.

With n items there are $n - 1$ degrees of freedom, for there are $n - 1$ comparisons involving x_1, viz. $x_1 - x_2, x_1 - x_3, \ldots x_1 - x_n$, and any other difference between x's can be expressed as the difference of two of these.

A simple example will show how t can be used to calculate confidence limits.

Example 13.3

Six independent observers estimate the melting-point of a certain substance as $354°$, $359°$, $362°$, $353°$, $356°$ and $358°$ C respectively. Find the 99 per cent confidence limits for the true value.

The mean of the six observations is $357°$ C, and this is the best estimate of the true melting-point. The six deviations are (in degrees) -3, $+2$, $+5$, -4, -1 and $+1$.

$$\therefore s^2 = \tfrac{1}{5}\{(-3)^2 + 2^2 + 5^2 + (-4)^2 + (-1)^2 + 1^2\}$$
$$= 11\cdot2$$

$$\therefore \text{SE of mean} = \frac{s}{\sqrt{6}}$$
$$= \sqrt{\frac{11\cdot2}{6}}$$
$$= \sqrt{1\cdot8667}$$
$$= 1\cdot37$$

For five degrees of freedom and $P = 0\cdot01$, $t = 4\cdot03$, so the 99 per cent confidence limits are

$$357 \pm 1\cdot37 \times 4\cdot03 = 357 \pm 5\cdot5 \text{ approximately}$$
$$= 351\cdot5° \text{ and } 362\cdot5°$$

Tests of Significance

Occasions sometimes arise when it is necessary to compare a statistic, such as a mean or proportion, with some specified value, or to compare two statistics with each other. For example, an experiment may be carried out to test whether a new fertilizer improves the yield of a crop, or whether a change in a chemical process increases the yield or improves the quality of the product. There will inevitably be differences in the results. The question is whether those differences are *statistically significant*, i.e. whether they indicate a real improvement in the process, etc., or whether they could reasonably be attributed to chance in the form of sampling errors.

If a difference as great as, or greater than, the observed difference could only occur by chance less than once in 100 times, it is said to be significant *at the 1 per cent level,* and similarly for other *levels of significance.* A result that is significant at the 5 per cent level, i.e. a result that could only have occurred by chance less than 5 times (but more than once) in 100 is generally described as "probably significant." The 1 per cent and 5 per cent levels have no intrinsic merit, but they have proved convenient in practice and have become fixed by convention.

Putting it another way, we carry out our calculations on the assumption that the effect we are looking for does not really exist, then checking to see whether the results are consistent with that assumption. (Statisticians call this assumption the *Null Hypothesis*, but do not let that worry you.) The level of significance is the risk we are willing to take of being wrong.

Example 13.4

A plant is producing large numbers of articles of which, on the average, 2 per cent are defective. In a random sample of 1,000 there are 3 per cent defective. Does this indicate a significant deterioration in the process?

The expected number of defectives is 2 per cent of 1,000, i.e. 20.

Then SE of number defective

$$= \sqrt{(1{,}000 \times 0{\cdot}02 \times 0{\cdot}98)}$$
$$= \sqrt{19{\cdot}6}$$
$$= 4{\cdot}43$$

The difference between actual and expected numbers is 10, which is a little over twice the standard error. This is significant at the 5 per cent level, i.e. there is probably a real deterioration in the process.

Single-Sided and Double-sided Tests

In Example 13.4 the test used was *double-sided,* i.e. the probability taken as the criterion of significance was the probability of the difference from the mean of 10 or more in either direction. Sometimes, however, it is only differences in one direction that are of any interest, and in such cases a *single-sided* test is appropriate. Thus, if a new drug is being tested for a certain disease, and judged by the proportion of patients cured, the level of significance will be the probability that an increase (not merely a difference) of the magnitude found could have occurred by chance.

Example 13.5

A process takes an average time of 35 minutes, with a standard deviation of 2·4 minutes. It is thought that a certain modification would reduce this time, and after being modified the process is repeated 12 times, giving an average time of 33·3 minutes. Is there any significant reduction in the time? (It is assumed that the variability is unchanged.)

$$\sigma = \text{SE of mean time} = \frac{2\cdot4}{\sqrt{12}} = 0\cdot693 \text{ min.}$$

$$x = \text{Expected mean—actual mean} = 35 - 33\cdot3 = 1\cdot7 \text{ min.}$$

$$\therefore \frac{x}{\sigma} = \frac{1\cdot7}{0\cdot693} = 2\cdot45$$

For a single-sided test, $P < 0\cdot01$, i.e. a difference of 2·45 or more *in the direction expected* would not occur by chance once in 100 times. It is therefore significant.

Table 13.2 shows the values of x/σ (the Normal Deviate) corresponding to various values of P for both single-sided and double-sided tests.

Table 13.2   Values of the Normal Deviate for single-sided and double-sided tests

P	Single-sided	Double-sided
0·1	1·282	1·645
0·05	1·645	1·960
0·02	2·054	2·326
0·01	2·326	2·576
0·001	3·090	3·291

Comparison of two Estimates

The last two examples dealt with cases in which one sample statistic was compared with a known value. When two sample statistics, both subject to sampling errors, are compared with each other the problem is slightly more complicated.

We have seen that if x and y are any two random variables or "statistics" with standard errors σ_x and σ_y respectively, the standard error d of their difference ($d = x - y$) is given by Formula 13.2a,

$$\sigma_d^2 = \sigma_x^2 + \sigma_y^2$$

In the particular (and usual) case in which the expected value of d is zero and we are testing to see whether the actual difference between two sample means or proportions could have occurred by chance, we calculate σ_d, divide it into d, and compare the results with the appropriate table.

(a) Comparison of two sample proportions

Suppose two samples of n_1 and n_2 items are being compared in respect of a certain attribute, and the proportions of successes in those samples are p_1 and p_2 respectively. Then the total number of successes in the combined sample is $n_1 p_1 + n_2 p_2$, and the overall proportion of successes is given by

$$p = \frac{n_1 p_1 + n_2 p_2}{n_1 + n_2}$$

On the assumption that the two samples were taken at random from the same population (i.e. on the Null Hypothesis), this value p is the best estimate of the true proportion in the population. The standard error of the observed proportion in a sample of n items is therefore, from (13.7),

$$\sqrt{\frac{pq}{n}}, \text{ where } q = 1 - p$$

So, from 13.2a and 13.7, the standard error σ_d of $p_1 - p_2$ is given by

$$\sigma_d{}^2 = \sigma_{p_1}{}^2 + \sigma_{p_2}{}^2 = \left(\frac{1}{n_1} + \frac{1}{n_2}\right) pq \qquad (13.9)$$

Then $p_1 - p_2$ can be compared with σ_d by a single-sided or double-sided test, whichever is appropriate.

Example 13.6

Out of 200 articles made on one machine, 22 per cent are faulty; out of 120 made on another machine, 30 per cent are faulty. Is there a significant difference between the two machines? (Note the assumption that if the difference is significant, it is due to the machines. Such assumptions are often implied in questions without being stated.)

The proportion of faulty articles in the combined batch of 320 is

$$\frac{44 + 36}{320} = \frac{1}{4}$$

So, from 13.9,

$$\sigma_d{}^2 = \left(\frac{1}{200} + \frac{1}{120}\right) \times \frac{1}{4} \times \frac{3}{4}$$

$$= \frac{1}{400}$$

$$\therefore \sigma_d = \frac{1}{20}, \quad \text{or 5 per cent.}$$

The actual difference of 8 per cent between the two samples is only 1·6 times its standard error, and a difference as great as this could occur by chance at least 1 in 10 times. The verdict is therefore "not proven." Note that we have not proved either that there is a real difference or that there is not. Probably, further experiment is indicated.

(b) *Comparison of two sample means*

Suppose we wish to compare the means $\bar{x}_1$ and $\bar{x}_2$ of two samples, of size n_1 and n_2 items respectively, to check whether they could reasonably have come from the same distribution. Let us suppose first that the standard deviation σ of that distribution is known.

From formula 13.5, the standard errors of $\bar{x}_1$ and $\bar{x}_2$ are respectively $\sigma/\sqrt{n_1}$ and $\sigma/\sqrt{n_2}$, and from formula 11.2a, the standard error of $\bar{x}_1 - \bar{x}_2$ is given by

$$\sigma_d{}^2 = \frac{\sigma^2}{n_1} + \frac{\sigma^2}{n_2} = \left(\frac{1}{n_1} + \frac{1}{n_2}\right)\sigma^2 \tag{13.10}$$

Example 13.7

Two machines both produce articles varying in weight with a standard deviation of 1·5 oz; one turns out 50 with a mean weight of 17 lb 2·1 oz and the other turns out 30 with a mean weight of 17 lb 2·9 oz. Is one machine biased with respect to the other? The standard error of the difference is, from (13.10),

$$\sigma_d = \sqrt{\left(\frac{1}{50} + \frac{1}{30}\right)} \times 1\cdot5$$

$$= 0\cdot346$$

The actual difference of 0·8 oz is 2·31 times its standard error, and the table of the Normal distribution shows that a difference as great as this could only occur by chance about once in fifty times. We conclude that one of the machines is probably biased.

If σ is not known, it must be estimated by pooling the sums of squares from the two samples. Strictly speaking, the two estimates of σ^2 should be compared for consistency, but that takes us beyond the scope of this chapter.

From first sample, expected value of

$$\Sigma(x - \bar{x}_1)^2 = (n_1 - 1)\sigma^2$$

and from second sample, expected value of

$$\Sigma(x - \bar{x}_2)^2 = (n_2 - 1)\sigma^2$$

The best estimate of σ^2 is therefore s^2, where

$$(n_1 + n_2 - 2)s^2 = \Sigma(x - \bar{x}_1)^2 + \Sigma(x - \bar{x}_2)^2$$

The standard error of $d(= \bar{x}_1 - \bar{x}_2)$ is then given by

$$s_d{}^2 = \left(\frac{1}{n_1} + \frac{1}{n_2}\right) s^2 \qquad (13.11)$$

If n_1 and n_2 are small, s^2 is uncertain and we must again use t instead of the Normal deviate. Since s^2 has been obtained with $n_1 + n_2 - 2$ degrees of freedom, this is the appropriate number to use with t.

Example 13.8
Seven eight-week old chickens reared on a special diet weigh 13, 16, 12, 17, 15, 15 and 16 ozs, and five chickens of similar age fed on ordinary diet weigh 9, 11, 15, 11 and 14 ozs. Is there sufficient evidence that the special diet has caused an increase in weight?

Denoting the two groups by suffixes 1 and 2 respectively, it is easily seen that

$$\bar{x}_1 = 15, \qquad \bar{x}_2 = 12, \qquad d = 3$$
$$\Sigma(x - \bar{x}_1)^2 = 19, \qquad \Sigma(x - \bar{x}_2)^2 = 24$$
$$\therefore\ 10s^2 = 19 + 24 = 43$$
$$s^2 = 4\cdot3$$
$$\therefore\ s_d{}^2 = (\tfrac{1}{7} + \tfrac{1}{5})4\cdot3 = 1\cdot474$$
$$s_d = 1\cdot214$$
$$\therefore\ t = \frac{3}{1\cdot214} = 2\cdot47$$

Since s^2 has been calculated with 10 degrees of freedom, we look up the corresponding values of t in Table 13.1 (page 139) and find that a value as great as $2\cdot23$ would be significant at the 5 per cent level. But be careful: that is for a double-sided test. We are only interested in differences in one direction. An *increase* in average weight of the amount found is significant at the $2\cdot5$ per cent level, i.e. it would occur by chance less than once in forty times. We conclude that the effect is probably significant.

Paired Comparisons
It is often possible to increase the sensitivity of a test by removing most of the variability from our data. In Example 13.8, for instance, we might have reduced our sum of squares and consequently our value of s^2 by pairing off chickens from the same brood and allocating one to the special diet and the other to the ordinary diet. The allocation must, of course, be done by some random process.

Example 13.9

In order to test a possible treatment for increasing the abrasion resistance of rubber, eight test-pieces are taken from different sheets and each test-piece is divided into two parts, one of which receives the treatment. All sixteen pieces are tested, with the following results—

Test-piece	Treated	Untreated
1	15	12
2	14	11
3	13	13
4	16	14
5	12	11
6	12	10
7	15	14
8	11	11

If these data were analysed as in Example 13.8, we should obtain $t = 1.82$ for $14df$, which is barely significant at the 5 per cent level. However, there is considerable variation between test-pieces, which can be removed by taking the differences between corresponding halves and analysing the eight figures so obtained.

Test-piece	Difference (x)
1	3
2	3
3	0
4	2
5	1
6	2
7	1
8	0

The fact that there is not a single negative difference is itself fairly conclusive (let us use our common sense where we can). However, to illustrate the method, we find that

$$\bar{x} = 1.5, \qquad \Sigma(x - \bar{x})^2 = 10$$

$$\therefore s^2 = \frac{10}{7}$$

and SE of $\bar{x}$

$$= \frac{s}{\sqrt{8}} = \sqrt{\frac{10}{56}}$$

Since we are testing whether $\bar{x}$ is significantly greater than (not merely different from) zero,

$$t = 1.5 \div \sqrt{\frac{10}{56}} = 3.55.$$

For 7 *df*, this is significant at the 1 per cent level—in fact it would be, even for a single-sided test. Thus, what we have lost in degrees of freedom we have more than gained in sensitivity, by reducing the standard error.

Standard Errors of Median and Standard Deviation

The standard error of the mean of a sample is very simple and does not depend on the form of the parent distribution. The standard error of the median is more complicated, but for a Normal parent distribution it is approximately $1.25\sigma/\sqrt{n}$, compared with $\sigma/\sqrt{n}$ for the mean. In other words, the mean gives a more reliable estimate of the central value than the median, provided that the distribution is Normal, or approximately so.

Sometimes it is necessary to compare standard deviations, e.g. when an attempt is made to reduce the variability of a process. The standard error of a standard deviation is approximately $\sigma/\sqrt{2n}$ *for a Normal population.*

The proofs of these two formulae will not be given here. Both formulae may be considerably modified when the parent distribution is far from Normal.

Concluding Remarks

Only the fringe of this subject can be touched in an elementary textbook, and it has only been possible to deal cursorily with the treatment of small samples. The reader who is sufficiently interested and equipped to read further should refer to more advanced books. Certain aspects of sampling methods will be discussed further in Chapter 15.

It only remains to sound a note of caution concerning tests of significance. The conventional 1 per cent and 5 per cent are useful so long as they are treated with discretion. An experimental result significant even at the 10 per cent level may justify changing a process if it supports an opinion already held for good reasons. On the other hand, the management will need something more convincing than a result that is "probably significant" before embarking on a large capital construction programme. Indeed, they will need to be satisfied that the improvement is great enough to justify the change.

This raises a point of great importance. A test of significance often tells nothing about the size of the effect being tested. A difference may be statistically significant but in practice insignificant. Confidence intervals are often more informative than tests of significance, because they show the probable size of the difference being investigated and at the same time indicate the degree of uncertainty attached to it.

Exercises

13.1 Give formulae for the following, explaining your notation and stating what assumptions have to be made in calculating and in interpreting them—

(a) Standard error of a mean.
(b) Standard error of a proportion.
(c) Standard error of a number of accidents.
(d) Standard error of a sum of two readings from the same population.
(e) Standard error of the difference between two readings from the same population. (IoS)

13.2 In order to find whether intensive national advertising of milk affected sales in a particular area, two systematic samples of families were taken from the milkroundsmen's books, one before and one after the campaign. With the help of means and standard deviations, determine whether the campaign was successful, ignoring all other factors influencing the pattern of milk sales.

Number of pints bought per week	Number of families in	
	Sample 1 Before the campaign	Sample 2 After the campaign
Up to 5	77	81
$5\frac{1}{2}$–$10\frac{1}{2}$	151	128
11 –16	213	194
$16\frac{1}{2}$–$21\frac{1}{2}$	254	295
22 –27	428	372
$27\frac{1}{2}$–$32\frac{1}{2}$	230	323
33 –38	92	146
$38\frac{1}{2}$ and over	45	79

(ICWA)

13.3 After corrosion tests, 42 of 536 metal components treated with Primer A and 91 of 759 components treated with Primer B showed signs of rusting. Test the hypothesis that Primer A is superior to Primer B as a rust inhibitor. (ICWA)

13.4 What is meant by the term "confidence limit"?

In a random sample of Manchester residents 410 people, out of 1,000 interviewed, looked at a certain television programme. Find the 95 per cent confidence limits for the percentage of all the residents who looked at the programme. If a sample of 500 Londoners showed that 52 per cent looked at the same programme, would you consider this evidence of a difference in viewing habits between the two cities? (IoS)

13.5 In a sample of 500 garages it was found that 170 sold tyres at prices below those recommended by the manufacturer.

(a) Estimate the percentage of all garages selling below list price.

(b) Calculate the 95 per cent confidence limits for this estimate and explain briefly what these mean.

(c) What size sample would have to be taken in order to estimate the percentage to within 2 per cent? (ICWA)

13.6 A certain large country has two political parties "Blues" and "Greens." Approximately 55 per cent of the electorate vote Blue and 45 per cent vote Green.

(a) It is required to estimate with 95 per cent confidence, percentages within ±1 unit. Find the sample size needed.

(b) Because of known differences in their traditional voting habits, you are now told that 75 per cent of the electorate is wage-earning and 25 per cent salaried or self-employed. In a sample of the size calculated in (a) above, only 72 per cent of the persons interviewed were found to be wage-earning. To what extent would this affect any conclusions drawn from the sample? What remedy would you propose for such a situation? (IoS)

13.7 Seeds of a certain vegetable are required by law to have a minimum germination rate of 80 per cent when they are sold. A seedsman performs a trial of 50 seeds and observes that 45 germinate. Is he justified in believing that his seeds are up to standard?

An inspector tests a further random sample of 50 seeds and finds that only 32 germinate. Is this sufficient evidence that the stock is substandard? Are the inspector's results consistent with those of the seedsman? (IoS)

13.8 Explain what is meant by the term "confidence interval."

A sample of 12 observations from a certain population gave the following total and sum of squares—

$$\Sigma x = 24, \qquad \Sigma x^2 = 92.$$

Derive a 95 per cent confidence interval for the population mean.

At a later date a further sample of 15 observations gave the total and sum of squares—

$$\Sigma y = 45, \qquad \Sigma y^2 = 241.$$

Stating any assumptions you make, test at the 5 per cent level of significance whether the population mean has changed. (IoS)

13.9 The time taken for a maintenance job on part of a textile process has been established to have a mean of 3 hrs 24 mins and a standard deviation of 12 minutes. How many tests would be required by a new method of maintenance to detect a difference of 9 minutes in average time for the job with 95 per cent confidence? The following eight times were recorded by a new method—

3 hrs 6 mins	3 hrs 12 mins
2 hrs 54 mins	3 hrs 0 mins
3 hrs 18 mins	3 hrs 24 mins
3 hrs 30 mins	3 hrs 12 mins

Obtain an estimate of the mean of the new method and calculate its 95 per cent confidence limits.

Is there any indication that the average time for the new method differs from the average time for the established method?   (IoS)

13.10 A large number of employees are engaged in an office on similar routine tasks. It has been suggested that a change in the equipment used will improve productivity. The efficiency of ten employees selected at random is measured with the old and the new equipment. The following data are then available—

| Employee | Efficiency rating | |
	Old equipment	New equipment
A	60	60
B	64	67
C	56	60
D	45	48
E	96	94
F	64	64
G	54	59
H	40	50
I	78	78
J	75	77

Do the data provide evidence of the superiority of the new equipment?   (IoS)

13.11 In a clinical comparison of the effectiveness of two aerosols, *A* and *B*, in the relief of bronchial asthma, ten patients were selected. Each patient was given *A* on one occasion and *B* on a different occasion, the order of administration being decided by a random procedure. On each occasion the patient's forced expiratory volume (FEV) was estimated before treatment and at intervals up to one hour after

treatment. The criterion of effectiveness was taken to be the maximum percentage increase in FEV compared with the pre-treatment
readings—

Patient number	Maximum percentage increase in FEV after	
	Aerosol A	Aerosol B
1	28	24
2	22	16
3	10	5
4	40	17
5	18	23
6	57	52
7	49	30
8	40	16
9	34	14
10	63	31

Is there any convincing evidence of a difference between the two
aerosols?   (IoS)

14

Regression and Correlation

THE previous few chapters have been concerned with the distribution of a single variable. The present chapter deals with the interrelations between two or more variables.

The Meaning of Regression

In the exact sciences like astronomy, chemistry and physics, it is possible to formulate laws connecting several quantities, e.g. the density, temperature and pressure of a gas, so that any one of these quantities can be determined (subject only to small experimental errors) from the others. In meteorology, prediction is much less certain, but it is still possible to forecast the weather with considerable success. In the living sciences like biology and agriculture, there are so many unknown factors at work that very different results may be obtained from what appear to be identical causes.

Even here, however, the law of averages holds good, e.g. big men and women tend to have big children, although the association is only partial. The adult sons of men below average height will also tend to be short, but on the whole they will be nearer the average height than their fathers, e.g. the sons of men 5 ft tall might, on the average, reach about 5 ft 4 in. It was this "regression" to the normal, noticed by Sir Francis Galton in his research on heredity, that gave the name "regression analysis" to this branch of statistics.

While, therefore, the son's height cannot be deduced from the father's, it is possible, knowing the father's height, to make a better forecast of the son's height than by simply taking the average of the population. In Chapter 13 it was shown how to estimate a parameter and find the precision of that estimate. In this chapter it will be shown how to find the best estimate of one variable partly dependent on another, knowing the value of the other, and to find the precision of the estimate.

Correlation

Two or more variables that tend to move in sympathy are said to be *correlated*. In an elementary textbook it is possible to consider only two variables. When high values of one are associated with high values of the other, as in the case of fathers' and sons' heights, they are said to be *directly* or *positively correlated*. When high values of one tend to accompany low values of the other, e.g. unemployment and labour turnover, they are *inversely* or *negatively correlated*.

There are of course degrees of correlation, i.e. two variables may be highly correlated or only slightly correlated. Thus, candidates' examination marks in two mathematics papers are likely to be highly correlated, but their marks in mathematics and art or woodwork will show little if any correlation. It will be shown later how the degree of correlation between two variables can be measured.

Scatter Diagrams

The existence of correlation can be shown graphically by means of a *scatter diagram*, in which each point corresponds to a pair of observations, one variable being plotted horizontally and the other

Table 14.1   Average duration of absence due to injury in various factories in 1969 and 1970

Factory	1969	1970
	Hours	Hours
A	232	205
B	206	175
C	252	273
D	101	122
E	154	176
F	224	236
G	158	217
H	321	324
I	153	151
J	74	46
K	154	109
L	170	183
M	303	226

vertically. Table 14.1 shows hypothetical data collected for different factories on the average duration of absence (in working hours) due to industrial injuries in successive years. Clearly there is wide variation between factories, but the scatter diagram shown in

Fig. 14.1 suggests that these averages are fairly consistent from year to year. From this one might infer that the severity of the injury depends largely on the type of work done in each factory.

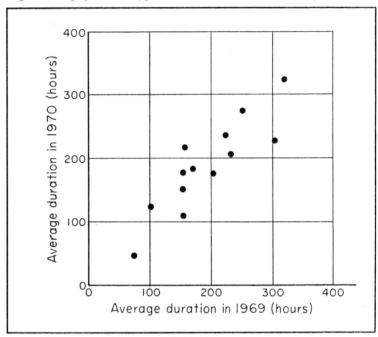

Fig. 14.1   Scatter diagram (from Table 14.1)

Curves and Lines of Regression

When there are many observations, the labour of plotting a scatter diagram becomes prohibitive, and it is better to compile a *correlation table*, which may be described as a two-way frequency distribution. An example of this is shown in Table 14.2, which has been adapted from a table in the Registrar-General's *Statistical Review for 1951*, the frequency of each "cell" being divided by 1,000 and rounded off to the nearest integer.

For each value or class-interval of one variable, the average value of the other variable may be calculated and plotted on a *regression diagram*. The points representing the average values of *y* corresponding to specified values of *x* will generally lie on or close to a smooth curve, called the *curve of regression of y on x*, and if, as often happens, this curve is almost a straight line, it is customary to fit a straight line by a method shortly to be described. This line is called the *line of regression of y on x* and its equation is called a *regression equation*.

Table 14.2 Correlation table showing combined ages at marriage of 353 married couples

Ages of husbands

		16–19	20–24	25–29	30–34	35–39	40–44	45–49	50–54	55–59	60–64	65–69	Total
	16–19	6	37	9	1	—	—	—	—	—	—	—	53
	20–24	3	95	57	13	3	1	—	—	—	—	—	172
	25–29	—	11	28	14	6	2	1	—	—	—	—	62
Ages of wives	30–34	—	1	6	8	5	3	1	1	—	—	—	25
	35–39	—	—	1	3	4	3	2	1	—	—	—	14
	40–44	—	—	—	1	2	3	2	1	1	—	—	10
	45–49	—	—	—	—	1	1	2	2	1	1	—	8
	50–54	—	—	—	—	—	—	1	1	1	1	—	4
	55–59	—	—	—	—	—	—	—	—	1	1	1	3
	60–64	—	—	—	—	—	—	—	—	—	1	1	2
	Total	9	144	101	40	21	13	9	6	4	4	2	353

The lines of regression of y on x, and of x on y, do not normally coincide; indeed, they will only do so if x and y are perfectly correlated. In the extreme case where x and y are completely independent, one regression line will be horizontal and the other vertical (ignoring sampling errors) because to every value of x (or y) will correspond the same mean value of y (or x).

Fig. 14.2 shows the regression of husband's age on wife's age and vice versa. The broken lines A and B would form smooth curves of regression if based on very large numbers of observations. Obviously there is a high correlation between the ages of husband and wife, as would be expected.

As a rule only one regression line or curve is required, because one variable y is generally regarded as dependent on the other variable x.

Regression Equations: The Method of Least Squares

It will be assumed in this section that the regression of y on x is linear, i.e. that it is sufficient to fit a straight line to the data. The same principles, however, can be extended to parabolas and other curves. The non-mathematically-minded student may omit this section as far as Equation (14.5).

Suppose there are n pairs of items

$$(x_1, y_1), (x_2, y_2) \ldots (x_n, y_n),$$

and that it is required to find the regression line of y on x.

It will save confusion if X and Y represent the variables and x and y denote observed values of X and Y. Let the required line be $Y = a + bX$, where a and b are constants to be determined. Then

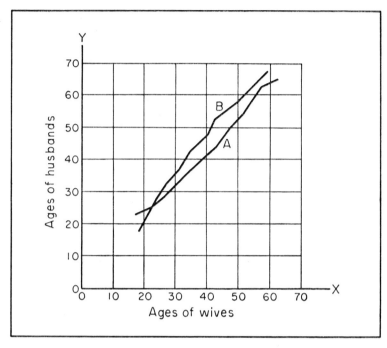

Fig. 14.2   Regression diagram illustrating regression of husband's age on wife's age and vice versa (from Table 14.2)

any observed value of Y, say y_r, may be regarded as the sum of (i) the estimate from the regression equation, viz. $a + bx_r$, and (ii) a residual $y_r - a - bx_r$. The Method of Least Squares consists in finding the values of the *regression coefficients a and b* that minimize the sum of squares of the residuals. This is equivalent to minimizing the standard error of an estimated value of y.

To simplify the mathematics, suppose both variables are measured from their means, so that

$$\Sigma x = \Sigma y = 0 \qquad (14.1)$$

Then $\Sigma(y - a - bx)^2$ must be minimum.

Differentiating partially with respect to a,

$$\Sigma(y - a - bx) = 0 \qquad (14.2)$$

$$\therefore \Sigma y - na - b\Sigma x = 0,$$

and from equations (14.1),

$$a = 0. \qquad (14.3)$$

Hence the regression line goes through the centroid of the points, and the sum of squares reduces to $\Sigma(y - bx)^2$.

Again, differentiating partially with respect to b,

$$\Sigma x(y - bx) = 0 \qquad (14.4)$$

$$\therefore b = \frac{\Sigma xy}{\Sigma x^2} \qquad (14.5)$$

The line of regression of Y on X is therefore—

$$Y = \frac{\Sigma xy}{\Sigma x^2} X, \qquad (14.6)$$

or, if X and Y are not measured from the means,

$$Y - \bar{y} = \frac{\Sigma(x - \bar{x})(y - \bar{y})}{\Sigma(x - \bar{x})^2}(X - \bar{x}) \qquad (14.7)$$

Similarly, the line of regression of X on Y is

$$X - \bar{x} = \frac{\Sigma(x - \bar{x})(y - \bar{y})}{\Sigma(y - \bar{y})^2}(Y - \bar{y}) \qquad (14.8)$$

If $\bar{x}$ and $\bar{y}$ are not convenient quantities, a short-cut method may be used for calculating the numerator of the regression coefficients similar to that found in Chapter 11 for $\Sigma(x - \bar{x})^2$.

$$\Sigma(x - \bar{x})(y - \bar{y}) = \Sigma xy - \bar{x}\Sigma y - \bar{y}\Sigma x + n\bar{x}\bar{y}$$

$$= \Sigma xy - \bar{x}(n\bar{y}) - \bar{y}(n\bar{x}) + n\bar{x}\bar{y}$$

$$= \Sigma xy - n\bar{x}\bar{y}$$

$$= \Sigma xy - \frac{1}{n}(\Sigma x)(\Sigma y) \qquad (14.9)$$

Example 14.1

Find the line of regression of Y (cinema admissions per head) on X (television licences per 100 population) from the following data for ten towns, covering one year.

Town	TV licences per 100 of population	Cinema admissions per head of population
A	12	12·8
B	39	10·9
C	16	13·7
D	30	12·4
E	43	8·8
F	17	13·3
G	47	10·1
H	15	11·6
I	34	10·6
J	22	10·4

Take trial means of 27 for X and 11 for Y and to simplify the arithmetic let

$$x = X - 27, \qquad y = 10\,(Y - 11)$$

The calculation is as follows—

Town	x	y	x^2	y^2	xy	$x + y$	$(x + y)^2$
A	−15	+18	225	324	−270	+ 3	9
B	+12	− 1	144	1	− 12	+11	121
C	−11	+27	121	729	−297	+16	256
D	+ 3	+14	9	196	+ 42	+17	289
E	+16	−22	256	484	−352	− 6	36
F	−10	+23	100	529	−230	+13	169
G	+20	− 9	400	81	−180	+11	121
H	−12	+ 6	144	36	− 72	− 6	36
I	+ 7	− 4	49	16	− 28	+ 3	9
J	− 5	− 6	25	36	+ 30	−11	121
Total	+ 5	+46	1,473	2,432	−1,369	+51	1,167

The last two columns are inserted as a useful check, since

$$\Sigma x + \Sigma y = \Sigma(x + y)$$

and

$$\Sigma x^2 + \Sigma y^2 + 2\Sigma xy = \Sigma(x + y)^2$$

Substituting,

$$5 + 46 = 51$$

and

$$1,473 + 2,432 + 2(-1,369) = 1,167$$

which confirms our arithmetic.

Then
$$\bar{x} = +0\cdot5,\ \bar{y} = +4\cdot6$$

$$\Sigma(x - \bar{x})^2 = 1473 - \frac{(+5)^2}{10} = 1470\cdot5$$

$$\Sigma(y - \bar{y})^2 = 2432 - \frac{(+46)^2}{10} = 2220\cdot4$$

$$\Sigma(x - \bar{x})(y - \bar{y}) = -1369 - \frac{(+5)(+46)}{10} = -1392$$

So the line of regression of y on x is

$$y - 4\cdot6 = - \frac{1392}{1470\cdot5}(x - 0\cdot5)$$

or, transforming to the original variables

$$Y - 11\cdot46 = - \frac{1392}{14705}(X - 27\cdot5)$$

which reduces to

$$Y = -0\cdot0947X + 14\cdot06$$

Fig. 14.3 shows the ten points and the regression line.

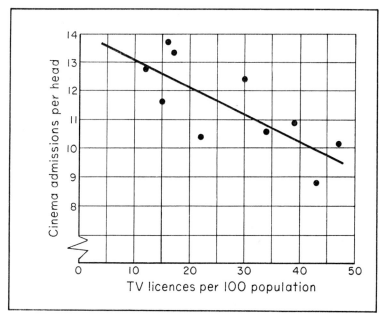

Fig. 14.3   Regression diagram illustrating example 14.1

Coefficient of Correlation

One object in fitting a regression line was to enable improved estimates to be given of the values of y corresponding to specified values of x, i.e. to reduce the standard error of such estimates. A comparison of Σy^2 with $\Sigma(y - bx)^2$, where x and y are measured from their means and b is given by Equation (14.5), will show how far this object has been achieved.

$$\Sigma(y - bx)^2 = \Sigma y^2 - 2b \Sigma xy + b^2 \Sigma x^2$$

$$= \Sigma y^2 - 2\left(\frac{\Sigma xy}{\Sigma x^2}\right)(\Sigma xy) + \left(\frac{\Sigma xy}{\Sigma x^2}\right)^2 (\Sigma x^2)$$

$$= \Sigma y^2 - \frac{(\Sigma xy)^2}{\Sigma x^2}$$

$$= (1 - r^2) \Sigma y^2, \tag{14.10}$$

where

$$r = \frac{\Sigma xy}{\sqrt{[(\Sigma x^2)(\Sigma y^2)]}} \tag{14.11}$$

This quantity r is called the *coefficient of correlation* between x and y. It is a suitable measure of correlation for the following reasons—

1. If there is no real correlation, the average value of y (ignoring sampling errors) is zero for every value of x, and $\Sigma xy = 0$, consequently $r = 0$.

2. If there is perfect linear correlation, every residual is zero, and from Equation (10), $r = \pm 1$. It is easily seen that $r = +1$ when correlation is direct and -1 when it is inverse.

3. Since $\Sigma(y - bx)^2$ cannot be negative, r cannot be numerically greater than 1.

Reverting to the original co-ordinates, Formula (14.11) becomes—

$$r = \frac{\Sigma(x - \bar{x})(y - \bar{y})}{\sqrt{[\{\Sigma(x - \bar{x})^2\}\{\Sigma(y - \bar{y})^2\}]}} \tag{14.11a}$$

Example 14.2

Find the coefficient of correlation between cinema admissions and TV licences in the data of Example 14.1.

The requisite sums of squares and products have already been calculated—in fact $\Sigma(y - \bar{y})^2$, was not needed for the regression line, but was included for convenience. From (14.11a),

$$r = \frac{-1392}{\sqrt{(1470 \cdot 5 \times 2220 \cdot 4)}} = -0 \cdot 77$$

In practice, we should hesitate to compute either the correlation coefficient or a regression line with only ten points. There does, however, seem to be a fairly strong negative correlation, and the regression line might help us to predict the cinema attendance of any town if we knew the number of TV licences and the size of population, provided we keep within the range of our observations. The regression line might not hold, for example, for values of X greater than 50.

The regression coefficients can be expressed simply by means of r. Putting

$$\Sigma(x - \bar{x})^2 = n\sigma_x^2 \text{ and } \Sigma(y - \bar{y})^2 = n\sigma_y^2,$$

$$r = \frac{\Sigma xy}{n\sigma_x\sigma_y} \tag{14.12}$$

and

$$b = \frac{\Sigma xy}{n\sigma_x^2} = r\frac{\sigma_y}{\sigma_x} \tag{14.13}$$

Hence the line of regression of y on x is—

$$Y - \bar{y} = r\frac{\sigma_y}{\sigma_x}(X - \bar{x}) \tag{14.14}$$

Similarly the line of regression of x on y is

$$X - \bar{x} = r\frac{\sigma_x}{\sigma_y}(Y - \bar{y}) \tag{14.15}$$

Significance of the Correlation Coefficient

Like other parameters, the correlation coefficient is subject to sampling errors, and it is customary to denote by ρ the true value, generally unknown, of which r is an estimate. The distribution of r is extremely complicated and an adequate treatment of it cannot be attempted here. Even when $\rho = 0$, i.e. when there is no real correlation, the distribution of r is by no means simple, but for large values of n the standard error of r may be taken as $1/\sqrt{n}$. Thus, if $n = 100$, an observed value of $0\cdot2$ is probably significant and one of $0\cdot3$ almost certainly significant.

Great care must be observed in judging whether or how far the correlation coefficient is significant in the everyday sense of the word. In the first place, it only measures linear correlation and is not suitable for data which, when plotted, form a parabola or similar curve. In the second place, correlation does not necessarily mean cause and effect. It is quite easy to find "nonsense correlations" between things that have no connexion whatever, except that they are both correlated with time or some other variable. Thus a rising divorce rate might be correlated with a falling mortality rate, but that would not mean that one caused the other.

Again, had we taken total cinema admissions and total number of TV licences in Example 14.1 instead of rates per head of population, we should have found them positively, not negatively, correlated because both would have been correlated with size of population.

Rank Correlation

It sometimes happens that the characteristics whose possible association is being investigated cannot be measured. Thus, a schoolmaster may want to know how far a boy's ability in games determines his popularity, but it would be extremely difficult to give either of these qualities a numerical value. However, it is fairly easy to place the boys in order for each of these qualities and to measure the agreement between the two rankings by means of *Spearman's rank correlation coefficient*.

The principle is the same as before, but the calculation is simplified because both x and y take the integral values $1, 2, \ldots n$.

$$\bar{x} = \bar{y} = \frac{n+1}{2}$$

$$\Sigma(x - \bar{x})^2 = \Sigma_1^n x^2 - n\bar{x}^2$$

$$= \frac{n(n+1)(2n+1)}{6} - n\left(\frac{n+1}{2}\right)^2$$

$$= \frac{n(n^2 - 1)}{12}$$

and similarly,

$$\Sigma(y - \bar{y})^2 = \frac{n(n^2 - 1)}{12}$$

$$\therefore r = \frac{12\Sigma(x - \bar{x})(y - \bar{y})}{n(n^2 - 1)}$$

This formula can be further simplified, since

$$2(x - \bar{x})(y - \bar{y}) = (x - \bar{x})^2 + (y - \bar{y})^2 - \{(x - \bar{x}) - (y - y)\}^2$$

$$= (x - \bar{x})^2 + (y - \bar{y})^2 - (x - y)^2$$

Putting $d = x - y$,

$$2\Sigma(x - \bar{x})(y - \bar{y}) = \Sigma(x - \bar{x})^2 + \Sigma(y - \bar{y})^2 - \Sigma d^2$$

$$= \frac{n(n^2 - 1)}{6} - \Sigma d^2$$

$$\therefore r = \frac{n(n^2 - 1) - 6\Sigma d^2}{n(n^2 - 1)}$$

$$= 1 - \frac{6\Sigma d^2}{n(n^2 - 1)} \tag{14.16}$$

Even when the items in question can be given numerical values, the rank correlation coefficient often saves much arithmetic with little loss of accuracy. Thus, in an inquiry whether labour turnover and absenteeism are correlated, turnover and absence rates could be calculated for a number of factories doing similar work, and the factories could be ranked in descending order for both turnover and sickness.

Example 14.3

Calculate Spearman's rank correlation coefficient for the data of Example 14.1.

Here x and y denote the ranks of the various towns, beginning with the highest value in each case.

Town	x	y	$d(= x - y)$	d^2
A	10	3	+7	49
B	3	6	−3	9
C	8	1	+7	49
D	5	4	+1	1
E	2	10	−8	64
F	7	2	+5	25
G	1	9	−8	64
H	9	5	+4	16
I	4	7	−3	9
J	6	8	−2	4
Total			0	290

The sign of d is immaterial, but signs have been inserted here to check that the sum of the differences is zero.

The rank correlation coefficient is, from (14.16),

$$1 - \frac{6 \times 290}{990} = -0.76$$

As usually happens, this agrees fairly well with the value of r found from the original values.

The rank correlation coefficient is often denoted by ρ, but as already stated, this symbol is now generally used for the true correlation coefficient of the population from which a sample is taken. Unfortunately, no alternative symbol has been generally accepted.

Exercises

14.1 What is the object of correlation analysis? Describe and illustrate:
(a) a scatter diagram; (b) a regression line; (c) perfect correlation; (d) negative correlation.   (ACCA)

14.2 Represent the following information regarding to passenger liners on a scatter chart. From an examination of your chart make your best estimate of the speed of a vessel of (a) 17,500 gross tons (b) 40,000 gross tons. Discuss the factors affecting the reliability of each of these estimates. Would you feel more confident about one of them than the other; if so, why?

Gross tonnage	Service speed (knots)	Gross tonnage	Service speed (knots)
20,900	20	13,200	18
14,500	22	21,100	22
6,050	16	14,300	18
8,800	$18\frac{1}{2}$	28,200	24
13,400	$15\frac{1}{2}$	29,600	22
25,700	$21\frac{1}{2}$	21,300	24
24,200	22	29,300	23
10,300	16	24,350	23
7,400	16	19,400	20
12,450	16	11,900	20

(IoT)

14.3 Calculate the line of best fit for estimating y from x for the following data—

y 13·0, 12·0, 16·5, 16·0, 17·5, 19·5, 19·0, 21·0

x 14·5, 17·0, 16·0, 17·5, 19·0, 20·5, 20·0, 20·5

Plot a scatter diagram and draw the calculated line on the graph.

(IoS)

14.4 Calculate the coefficient of correlation for the following data:

Expenditure on trading stamps £	Sales of petrol £
10,000	300,000
25,000	500,000
20,000	350,000
30,000	450,000
15,000	400,000

Explain the meaning of the term "coefficient of correlation" and state what its value indicates in this example.   (ACCA)

14.5 A manufacturer is considering the introduction of an incentive bonus scheme, but fears that it may lead to an increase in the proportion of rejects produced.
The following figures are the results of a sample of shifts worked over the past month

Production of operative per shift	Percentage of rejects
240	1
450	2
690	1
710	3
730	4
790	4
810	5
890	6
1,150	6
1,370	7

(a) Calculate a correlation coefficient for these figures.

(b) How would you use it to assess the likely results of such a bonus scheme?

(c) What reservation would you make in using the coefficient for this purpose?   (ICWA)

14.6 A destructive test on a range of metal castings has shown a fairly close correlation between measurement x and strength y.

Sample	Measurement x	Strength y
a	0·6	10
b	0·9	13
c	1·8	18
d	2·3	21
e	1·7	24
f	2·8	26
g	2·4	27
h	2·8	28
i	3·3	31
j	3·1	34
k	3·8	36
l	4·0	40

Using these results, compile a formula for estimating strength y from measurement x and predict the strength of castings with the following measurements of x: 1·0; 2·2; 2·5. (ICWA)

14.7 The values of $x = \log_{10}$ (mean temperature of the air at sea level, degrees F) and $y = \log_{10}$ (death rate per thousand, persons aged 85 and over, all causes) are given below for the years 1952 to 1966. Calculate the regression coefficient of y on x and hence deduce the relationship between the death rate for persons aged 85 and over and the mean air temperature.

Year	x	y	Year	x	y
1952	1·69	2·36	1960	1·70	2·34
1953	1·71	2·28	1961	1·71	2·36
1954	1·70	2·35	1962	1·69	2·36
1955	1·70	2·37	1963	1·68	2·37
1956	1·69	2·37	1964	1·70	2·31
1957	1·71	2·32	1965	1·69	2·32
1958	1·70	2·35	1966	1·70	2·34
1959	1·71	2·35			

(IoS)

14.8 The research department of your company has tested seven types of valve to see how efficiently they performed. The results, together with the prices of the valves are given below

Valve	Price £	Ranking by performance
A	132	2
B	98	6
C	117	1
D	89	5
E	145	3
F	100	4
G	85	7

Show by calculating a rank correlation coefficient whether in this instance price is a good guide to performance.   (ICWA)

14.9 Write down an expression for the rank correlation coefficient between two sets of rankings (x_i, y_i), $i = 1, 2, \ldots n$.

The following table gives the order in which two judges X and Y placed eight contestants in an essay competition

	A	B	C	D	E	F	G	H
X	1	5	3	4	2	7	6	8
Y	3	7	6	1	8	4	2	5

Calculate the coefficient of rank correlation and state any conclusions you draw from it.   (IoS)

15

Some Industrial Applications of Statistics

THIS chapter may almost be regarded as a continuation of Chapter 13, as it deals principally with certain applications of sampling theory to industrial problems. They are made necessary by the fact of variability in processes and experiments.

Variation in Industrial Processes

No matter how good the equipment or how skilful the operative, a number of articles produced by the same process will never be exactly alike, although with imperfect measuring instruments they may appear so. There will always be that slight variation from one to another, as if the material were endowed with an element of free will.

The causes of variation, which are many, may be divided into *assignable causes*, such as mechanical faults or abnormalities in the raw material or in the reaction conditions of a chemical process, which can be identified and eliminated, and *chance causes* which are either unknown or impossible to eliminate. When only chance causes are operating, the process is said to be *under control*. The data recorded will then vary about the *process average*, generally denoted by $\bar{X}$, with a standard deviation σ measuring the inherent variability of the process.

Control Charts for Variables

A useful device for keeping a check on the efficiency of a process is the *control chart*, on which observations are plotted, generally at regular intervals of time. The use of control charts for this purpose is called *Quality Control* or, to give it its full title, *Statistical Quality Control*.

The chart contains horizontal lines representing (i) the process average, and (ii) *control limits* outside which only a small specified proportion of observations will fall when the process is under control. It is customary to show two pairs of control limits, the *inner control limits* containing all but 5 per cent of such observations and the *outer control limits* all but 0·2 per cent. As with levels of significance, experience shows that these percentages are convenient as a rough guide.

Fig. 15.1 shows a typical control chart for single observations with

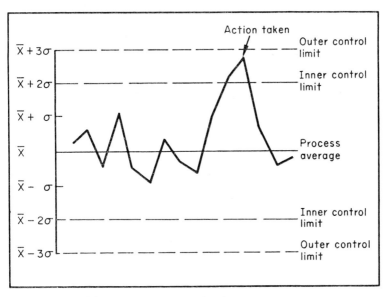

Fig. 15.1   Control chart for single observations

a process average $\bar{X}$ and a standard deviation (when the process is under control) of σ. The observations may be assumed Normally distributed, so the inner control limits are fixed at $\bar{X} \pm 1·96\sigma$ and the outer control limits at $\bar{X} \pm 3·09\sigma$, as these pairs of values contain respectively 95 per cent and 99·8 per cent of the Normal distribution.

The arrow in the figure shows where the process appeared to be going out of control and action was taken to correct it.

It is unusual to plot single observations, the general practice being to take a sample of five or ten observations at a time, so that both the average and the variability can be checked. As remarked in previous chapters, the most efficient measure of dispersion is the standard deviation, and in a single once-for-all check this would be the measure to calculate, but when samples are taken at frequent intervals such checks need not be so rigorous, and for ease and simplicity the range is generally used instead of the standard deviation.

Unfortunately, the distribution of the sample range w is not Normal; indeed it is too complicated to discuss here, but the mean and control limits have been tabulated for various values of n, the sample size. For $n = 5$, the mean range (denoted by $\bar{w}$) is about $2 \cdot 33\sigma$, the inner control limits are $0 \cdot 85\sigma$ and $4 \cdot 20\sigma$, and the outer control limits are $0 \cdot 37\sigma$ and $5 \cdot 48\sigma$.

Example 15.1

A machine is turning out steel rods with a process average of 6 inches and a standard deviation of $0 \cdot 005$ inch. Construct control charts for the mean and range of the lengths of samples of five rods.

The standard error of the mean of five items is

$$\frac{0 \cdot 005}{\sqrt{5}} = 0 \cdot 00224 \text{ in.}$$

Hence the inner control limits are

$$6 \pm 1 \cdot 96 \times 0 \cdot 00224 = 6 \pm 0 \cdot 0044 \text{ in.}$$

and the outer control limits are

$$6 \pm 3 \cdot 09 \times 0 \cdot 00224 = 6 \pm 0 \cdot 0069 \text{ in.}$$

The control chart for the mean is therefore as shown in Fig. 15.2.
For the range, $\bar{w} = 2 \cdot 33 \times 0 \cdot 005 = 0 \cdot 0116$ in.,

the inner control limits are

$$0 \cdot 85 \times 0 \cdot 005 = 0 \cdot 0042 \text{ and } 4 \cdot 2 \times 0 \cdot 005 = 0 \cdot 0210 \text{ in.,}$$

and the outer control limits are

$$0 \cdot 37 \times 0 \cdot 005 = 0 \cdot 0018 \text{ and } 5 \cdot 48 \times 0 \cdot 005 = 0 \cdot 0274 \text{ in.}$$

Fig. 15.3 shows the resulting control chart.

The control limits are only a guide, not a substitute for common sense. For example, two or three consecutive points just inside the inner control limits would suggest that the process needed investigating. Whether it were actually stopped for inspection would depend on the nature of the process, the cost of interrupting it, and so on.

It is not generally necessary to insert the lower control limits for the range, since reduced variation is a cause for satisfaction, not for concern. They may, however, be useful if an attempt is being made to find the causes of variation or to reduce the variability.

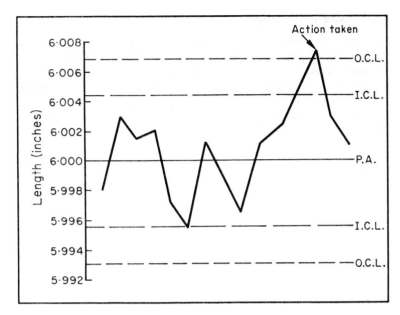

Fig. 15.2   Control chart for the mean length of a sample of five rods

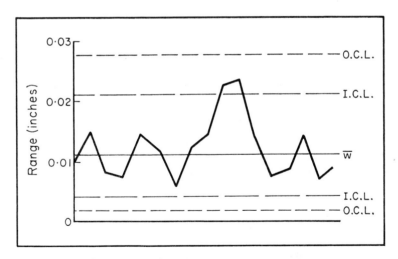

Fig. 15.3   Control chart for the range of the lengths of a sample of five rods

Control Charts for Attributes

The control charts described above are suitable for substances like coal, glass and fibres which can be measured in terms of one or more continuous variables such as density, length, tensile strength or the percentage of some ingredient; but there are manufactured articles like lamps, valves and sacks that cannot be measured in this way but are either good or bad. Such things can be checked only by taking fairly large samples and finding the proportion defective.

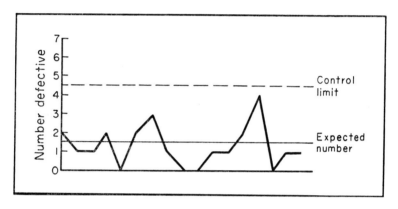

Fig. 15.4   Control chart for numbers of defectives in samples of fifty with a process average of 3 per cent defective

Suppose, for example, that a machine is producing large quantities of small articles of which, under normal conditions, 3 per cent are defective, and that samples of fifty are tested from time to time. Then the average number of defectives in a sample will be $1\frac{1}{2}$ and the actual number of defectives will be distributed in a Poisson distribution with $m = 1\cdot5$ (page 128). The numbers observed in samples of fifty can be plotted on a control chart as in Fig. 15.4.

One difficulty here is that no control limit can be fixed such that exactly 5 per cent of the observations are above it and 95 per cent below. In this example nearly 2 per cent of the samples will contain five or more defectives, and about $6\frac{1}{2}$ per cent will contain four or more. Accordingly a control line has been drawn between four and five, and the probability of a point lying above that line by chance is about $0\cdot02$.

If the 5 per cent confidence level is particularly desired, it can be obtained by finding a suitable sample size.

Cumulative Sum Charts

A useful alternative to the type of chart described above is the *cumulative sum chart*, in which the item plotted is not the number of defectives, say, in each individual sample, but the total number of defectives found in all samples taken to date, starting at a given time and continuing for a long period. Fig. 15.5 shows such a chart for the same data as were depicted in Fig. 15.4.

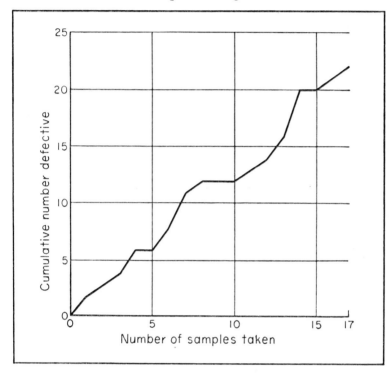

Fig. 15.5   Cumulative sum chart for date of Fig. 15.4

Over a fairly long period, the slope of the chart will provide an estimate of the average quality of the product, and a marked change in the slope will indicate a change in quality. These charts can also be used with continuous variables, the quantity cumulated being not the actual length, etc., but the deviation of that variable from the process average or standard value.

Advantages of Control Charts

The great advantage of control charts is that they enable a continuous watch to be kept on the process, thereby saving wastage of product. Furthermore, they provide a permanent record of performance to which customers can be referred to satisfy them that the product is of good quality. They also have a psychological effect in keeping operatives alert for faults and giving them a greater interest in their work. Control charts can be applied to many other fields besides industrial processes, e.g. accidents, typing errors, sickness rates and so on.

Acceptance Sampling

Acceptance sampling generally means the use of sampling methods to check the quality of purchased materials, but it can be applied to auditing of accounts and invoices and other forms of checking. It may be a single test for a particular *batch* or *lot*, or it may be a series of checks of successive lots, in which case the control chart technique can be used.

The difficulty with single lots is that unless the sample is very large it may give quite a misleading result. The tester must therefore take a certain risk (*consumer's risk*) of accepting an unsatisfactory batch, e.g. one containing 5 per cent or more defectives, and also a risk (*producer's risk*) of rejecting a batch that is actually good, e.g. containing less than 1 per cent defective. The actual quantities chosen are largely a matter of judgment.

Suppose, for example, that a purchaser wishes to be 95 per cent confident that a particular lot does not contain more than 2 per cent defective. Clearly, to give this degree of confidence, the percentage defective in the sample must be considerably less than 2 per cent. Thus, a sample of 150 from a lot with 2 per cent defective would contain only one defective about once in five times, so it is easy for a fairly large sample containing less than 1 per cent defective to mask a proportion of 2 per cent defective in the lot.

Consider the following plan—

(*a*) Take a sample of 150 items and, if no defectives are found, accept the lot. Otherwise, reject it.

This plan would ensure that 95 per cent of all lots with 2 per cent defective were rejected, but it would also result in the rejection of many very good lots simply because the rare defective happened to occur in the sample. For example, it would reject 53 per cent of all lots containing only $\frac{1}{2}$ per cent defective. Clearly, a plan that allows no defectives whatever is not sufficiently discriminating.

Now consider a slightly more selective plan—

(*b*) Take a sample of 240 items, and accept the lot if there is not more than one defective. If there are two or more, reject it.

This plan would also reject 95 per cent of all lots containing 2 per cent defective, but it would reject only 34 per cent of those containing ½ per cent defective. For a really selective plan it would be necessary to take a much larger sample.

Plans (a) and (b) are examples of *single sampling*, but *double sampling*, or sampling in two stages, is often more efficient and economical. The essence of double sampling is that if the first sample is indecisive, the result is decided on a second. On the average, this reduces the total amount of sampling, particularly where the quality of a lot is likely to be either very good or very bad.

The principle of double sampling can be extended to *sequential sampling*, in which observations are taken one at a time until a decision can be reached on the accumulated data, no sample size being fixed beforehand. After every observation a decision is taken either to accept the batch, or to reject it, or to continue sampling. Sequential sampling requires, on the average, much smaller samples than equivalent single or double sampling schemes, but the theory is complicated and sequential schemes are more difficult for an unskilled person to operate.

The words "accept" and "reject" have been used in a general sense, but rejection does not necessarily mean that the product is wasted. It may mean inspecting every item and replacing faulty ones. This is called *rectifying inspection*. Alternatively, the product may be sold at a reduced price.

Acceptance sampling usually requires large samples to give a reasonable degree of confidence in the quality of individual batches, and it is generally more economic and more satisfactory to keep a general check on quality by means of control charts, or to know that the supplier is doing so. In all sampling schemes it is necessary to balance the cost of inspection against the cost of occasional wrong decisions or failure to detect defective products. The ideal, in fact, is to minimize the average total cost in the long run.

Operational Research

Operational research is the application of scientific method, including statistical method, to the various operations of industry, commerce, warfare, etc. In industry this covers not only the manufacturing processes but such matters as stock control and distribution. In fact, these problems cannot be considered most efficiently in isolation. Stock policy, for example, will react on production planning, and both will depend on good sales forecasting. It will be convenient, however, to begin by considering the problem of optimum stocks.

Since demand is generally irregular and, to a large extent, un-predictable, it is often necessary to keep large stocks of finished goods in order to be able to meet any reasonable demand. If stocks are too low they will run out frequently and the manufacturer will lose business. On the other hand, if stocks are so high that it will hardly ever be necessary to refuse or delay an order, the cost of holding them may be prohibitive. Besides, some products may deteriorate or become obsolete if kept too long. It is therefore necessary to find the best compromise, and to balance the cost of holding stocks against the cost of occasionally being unable to fulfil an order promptly. Similarly with stocks of raw materials, it is necessary to balance the cost of storage against the cost of occasionally running out of materials and holding up production, or having to take emergency measures to obtain materials quickly.

Very often the problem can be solved by a *simulation study* based on past experience, on the assumption that the pattern of past demand can be taken as a fair guide to the future. The procedure is to take several years' data of, say, weekly or monthly demands, to con-struct a large number of possible sequences of orders by random sampling from those data and, assuming various levels of safety stock, to ascertain how often stocks would run out and what the total cost would be in each case, and then choose the safety stock for which the total cost is least. This technique, popularly called "Monte Carlo methods," can be used very effectively with the aid of an electronic computer.

The same technique can be employed with production planning. For example, in a firm making a wide range of products, e.g. varieties of paint, it may be a constant problem what to make next, and in what quantity. Probably the plant has to be cleaned out after one batch has been made before the next batch, of a different variety, can be started. It would be uneconomic either to make very small batches or to make such large batches as to create excessive stocks. Again the best compromise has to be found, and this problem, together with that of optimum stocks, can be programmed for a computer.

A rather similar problem is that of the best distribution of a product from a number of factories or depots to a large number of customers, given various transport costs, the total demand of each customer, the total capacity of each factory, etc. There are many other problems occurring in operational research, such as the right number of machines for one operative to tend when the machines break down at irregular intervals, and the best form of telephone or teleprinter network connecting the works and offices of a large company. The development of electronic computers has greatly facilitated the solution of such problems.

Statistical Methods in Industrial Research

In the research laboratory there is scope for statistical methods to deal with the uncertainty caused by experimental errors, instrumental errors, variations in materials, etc. Frequently, in an investigation to compare two methods or materials, say, or to estimate some physical quantity, one experiment is not enough. Too many would be wasteful, but too few would mean too great a risk of drawing wrong conclusions. Statistical methods help to determine the right number of experiments to make, and to assess the precision of the result.

It often happens that the yield or quality of a product depends on several factors, like temperature, reaction time and strength of solution. If these factors were independent of one another, it would be possible to determine first the optimum temperature, then the optimum reaction time, and so on. In practice, the problem is often complicated by interactions between the various factors. For example, the best temperature or pressure for one strength or reaction time may not be the best for another. Frequently the factors are so numerous, and can be varied in so many ways, that it is impossible or uneconomic to test every possible combination of them. It is necessary then to design a limited number of experiments in such a way as to extract the maximum amount of information at the minimum cost in time and money. Such *factorial experiments* can be designed, and the results analysed, by the aid of statistical methods.

This chapter has only touched the fringe of a vast subject which is still developing rapidly. For further information the reader should refer to more specialized books on industrial statistics, quality control, design of experiments, etc.; but perhaps enough has been said to show that statistics is neither a tool for the amateur nor a toy of the classroom, but a subject of great practical importance to industry as well as to the natural and social sciences.

Suggestions for further reading

Davies, O. L. (Ed.), *Statistical Methods in Research and Production* (Oliver and Boyd)

Davies, O. L. (Ed.), *Design and Analysis of Industrial Experiments* (Oliver and Boyd)

Duckworth, W. E., *A Guide to Operational Research* (Methuen)

Finney, D. J., *An Introduction to the Theory of Experimental Design* (University of Chicago Press)

Huitson, A. and Keen, J., *Essentials of Quality Control* (Heinemann)

Makower, M. S. and Williamson, E., *Teach Yourself Operational Research* (English Universities Press)

MOORE, P. G. and EDWARDS, D. E., *Standard Statistical Calculations* (Pitman)
PARADINE, C. G. and RIVETT, B. H. P., *Statistics for Technologists* (English Universities Press)
WETHERILL, G. B., *Sampling Inspection and Quality Control* (Methuen)

Exercises

15.1 Reels of insulated wire are made such that the lengths are Normally distributed with a mean length of 1,000 yds and a standard deviation of 10 yds. Two of these reels are selected at random and wound together to make a twin flex package. When one of the reels contributing to this package runs out, the wire remaining on the other package is regarded as waste. What average waste may be expected?

It is decided that a check will be kept on the amount of waste by actually measuring the waste from packages sampled at fairly frequent intervals. Set up a control chart and design record forms to do this.   (IoS)

15.2 The following are weights of medicinal tablets which are nominally 10 grains. Ten samples of 5 tablets have been taken at hourly intervals from the manufacturing process.

10·1	10·0	10·1	9·8	9·9	10·0	10·0	9·7	9·9	10·0
10·0	9·8	9·9	10·2	9·9	10·0	10·1	9·9	9·8	10·2
10·0	9·9	10·1	10·1	10·0	10·0	9·4	10·2	10·0	10·0
10·1	10·2	9·9	9·9	10·0	10·1	10·0	10·0	10·1	9·7
10·3	10·0	10·1	9·9	10·0	10·1	10·2	9·8	9·8	10·0

(*a*) Construct and plot suitable Quality Control Charts for this process.

(*b*) State and explain five criteria for detecting any changes which might occur in the manufacturing process.   (IoS)

15.3 The mean weight of a large loaf produced at a bakery was found in a series of quality control checks to be 30 ounces, with a standard deviation of 1 ounce.

Sample batches of 4 loaves were then taken from the travelling oven at 2-hourly intervals and the following are the averages of the last 6 consecutive sample batches:

29·6, 29·4, 29·9, 30·2, 30·3, 30·1

(*a*) Draw up a control chart for these averages and enter the figures on the chart.

(*b*) What proportion of loaves would you expect to fall below the legal minimum weight of 28 ounces?   (ICWA)

15.4 A machine makes bolts in batches of 1,024. Twenty consecutive batches were tested and the number of defective bolts in each is given below—

Batch	Number of defective bolts	Batch	Number of defective bolts
1	2	11	2
2	3	12	1
3	7	13	5
4	1	14	4
5	0	15	3
6	4	16	12
7	8	17	8
8	7	18	1
9	1	19	2
10	3	20	2

(i) Draw a control chart for the process showing control limits at any two levels you consider significant.

(ii) Enter the results shown above on to your chart and comment on the results. (IoS)

15.5 For a batch process making ceramic parts in large batches of approximately equal size, the percentage wastage is estimated by weighing the defective parts and quoting this weight as a percentage of the average weight of a batch. The 20 percentages tabled below are the percentage wastage in each of twenty successive batches.

Batch number	1	2	3	4	5	6	7	8	9	10
Per cent wastage	5·2	4·2	2·4	6·4	6·0	5·2	2·8	7·8	4·6	2·8
Batch number	11	12	13	14	15	16	17	18	19	20
Per cent wastage	3·4	5·3	2·9	4·6	2·5	6·5	4·1	7·3	3·5	7·5

Construct a control chart including and employing these first 20 observations, and discuss whether the process is in control.

(IoS)

15.6 The following figures were obtained from a series of quality control checks on engineering components, six items being measured at hourly intervals.

Time	9 am	10 am	11 am	12 noon	1 pm	2 pm
Sizes	119·0	120·8	119·7	119·9	118·5	119·7
	120·9	119·8	120·1	119·2	119·6	118·9
	119·0	118·9	120·9	121·1	120·2	123·1
	121·3	119·2	118·0	120·3	122·3	121·2
	120·0	120·5	118·6	122·6	122·9	118·0
	121·0	120·2	120·3	119·3	120·1	119·7

Depict these data on mean and range charts, using the following limits—

For the mean chart: Inner limits 119·5 and 120·5
                         Outer limits 119·2 and 120·8
For the range chart: Inner limit 3·5
                         Outer limit 4·5
Comment on the results.   (ICWA)

15.7 A certain assembly operation must be performed in a temperature- and humidity-controlled, dust-free atmosphere. To check this, an air sample is taken at hourly intervals and the dust particles above the critical size in a fixed volume of air are counted. The *twenty-five* counts taken during the first complete day, midnight to midnight, of the air-sampling system were: 0, 1, 1, 2, 3; 0, 1, 0, 0, 2; 2, 1, 1, 2, 0; 0, 3, 4, 0, 0; 0, 1, 0, 0, 1.

Construct an attribute control chart for the air-sampling and state the assumptions used in constructing the chart. Do these samples indicate a stable level for the dust content of the air?   (IoS)

15.8 The following numbers of defectives were obtained in samples of 50, taken at regular intervals—

Sample number	Number of defectives	Sample number	Number of defectives
1	10	11	8
2	5	12	14
3	19	13	5
4	16	14	12
5	14	15	17
6	18	16	10
7	7	17	14
8	2	18	17
9	11	19	21
10	8	20	24

Construct a control chart and comment as necessary.   (IoS)

15.9 Batches of 100 components were taken at fixed intervals of time from a production line and tested. The following figures are the number of components found to be defective in each of the batches:

4 3 2 0 7 5
2 4 1 3 4 6

(*a*) Calculate the mean proportion defective.

(*b*) Draw up a quality control chart for samples of 200 of the components, such that the inner limit is equal to the mean plus one standard error and the outer limit is equal to the mean plus two standard errors.

(*c*) Could your chart be used for batches of 50 components? If not, why not?   (ICWA)

16

Index Numbers

MOST readers will be familiar with the idea of an *index number* as a device for estimating trends in prices, wages, production and other economic variables. There is nothing abstruse about an index number. In its simplest form it represents a special case of an average, generally a weighted average, compiled from a sample of items judged to be representative of the whole.

Suppose an index of commodity prices is required. There are many commodities, with prices expressed in different units, but there is usually a tendency for prices of allied commodities to rise or fall together. Imagine the price of each commodity plotted on a graph in the form of daily or weekly quotations, or monthly or annual averages. Two or more such prices can be represented on the same graph if they can be reduced to the same units or dimensions, and a standard way of doing this is to express the current price as a percentage of the corresponding price at some fixed date in the past, called the *base date*, or of the average price for some given period, the *base period*. This percentage, known as a *price relative*, is independent of the units in which the price is quoted. The resulting graph will show a set of irregular lines fanning out from a single point corresponding to the base date or period, when each price relative is equal to 100. Unless these lines are very erratic, however, it should be possible to detect a general trend in the graph as a whole, corresponding to the movement of prices in general which it is the function of the index number to measure.

If the index number is obtained as an average of price relatives (as will be shown below, that is not the only way), it follows that the value of the index for the base date or period is also 100. In

181

algebraic formulae, however, it will be convenient to omit the factor 100. The weights (if any) allotted to the various items will depend on their relative importance, and since the importance of a commodity varies with time, place and the means and habits of those who use or trade in it, it is clear that there is no single solution to the problem, in the sense that one solution is right and all others are wrong. Fortunately, however, at least over short periods, solutions reached by different methods and with different weighting systems seldom differ by more than a few points.

The weights may either be fixed or vary from year to year, or even from month to month. When fixed weights are used, they often represent sales or similar quantities relating to a particular date or period, called the *weights base*. If, as generally happens, the weights base coincides with the *date base* (the base date or period defined above), the terms "base date" and "base period" may be used without ambiguity. If the weights base and date base differ, the terms "base date" and "base period" are taken to refer to the date base. If the weights are arbitrary or variable, there is, of course, no weights base.

Unless otherwise stated it will be assumed that the weights base (when it exists) is also the date base. It will also be convenient to take the base period as a year and to refer to index numbers of prices. In general, however, the principles and methods described in this chapter apply equally to indices of sales, production, wage-rates, etc. and to other base periods.

Choice of Items

The first, and probably the biggest, problem is to decide what items should be included in the index. They should be (1) relevant, (2) representative, (3) reliable, and (4) comparable over a period of time.

1. Items must be relevant, e.g. a producers' price index should be composed of wholesale prices, and a cost of living index should comprise retail prices together with rent, rates, etc. One should not use an index of retail prices to estimate changes in construction costs, or an index of wage rates of manual workers to estimate changes in the level of salaries of business executives.

2. An index number is generally based on a sample. For example, it would be impossible to include in an index of food prices every conceivable variety of food. It is desirable, therefore, that the items chosen should be adequate in number and importance, and representative of the whole. Generally, the more items there are, the more reliable will the sample be, and the less likely to be unduly affected by abnormal movements in one or two commodities. After a certain point, however, greater precision is outweighed by the expense and labour of collecting additional information.

In some cases, e.g. where a price index is being compiled for the products of a firm or of a fairly simple industry, it may be possible to cover the bulk of them with relatively few items. Often, however, a commodity is sold in many different sizes, grades, packages, etc., and under various contracts, and it may be necessary to select some specified terms of sale as giving a representative price.

3. As has been said before (it is worth repeating), it is essential for the basic data to be as accurate as possible. Preference should be given to items for which reliable quotations can be obtained with confidence. No theoretical refinements in formulae or methods of computation can compensate for serious deficiencies in the data.

4. One of the greatest difficulties is variation in quality. Very often it is only possible to obtain an average price for a commodity like paint, which may vary tremendously in composition and quality. Similarly with iron and steel and other metals, where a change of average price may merely reflect changes in the type, grade, etc. of the product. In such cases we have to be content with a unit value index, derived not from prices for specified grades, contract terms, etc., but from average values per ton or other unit of what may be very mixed bags of products. One simply hopes that changes of internal pattern average out fairly well.

It may also be necessary to exclude items where classification has changed or because data are not available for both base year and current year. A special case of this is products that were not known or not fully developed in the base year. Sometimes a new product is given a fictitious base-year price, but this is not very satisfactory. Another way of introducing new products is the chain base method (see page 192).

Sources of Data

When the items composing the "basket" have been decided, the next question is where the figures are coming from. Sometimes price quotations can be obtained from trade journals or from leading firms of dealers in the commodities in question, or they may be based on average import or export prices. Failing that, it may be necessary to take representative prices as reported by local observers (see Chapter 19).

Production data may be obtained similarly from published sources, trade associations or (by the Government) from producing firms.

The Base Period

The base period chosen is generally a year, and a fairly "normal" year for preference. If there is any difficulty in selecting a normal year, the average of a series of years can be taken. Thus, *The Statist*

Wholesale Price Index Numbers were for many years based on the period 1867 to 1877, and several American indices are based on a period of three years.

For indices of production, sales or other quantities, it is nearly always necessary to take at least a year as base period, since any shorter period might be unrepresentative and reflect seasonal movements; variation in the length of the months must also be allowed for. For many prices, wages rates, etc., it may be sufficient to take a particular day or the average for a single month. For example, the Board of Trade Index Numbers of Wholesale Prices were at one time based on 30 June 1949.

Progressive Changes in Index Numbers

For various reasons, most index numbers are subject to revision from time to time. Major changes may be due to far-reaching changes in conditions, habits and standards of life reflected in the pattern of expenditure, or to new ideas and techniques resulting in the adoption of new formulae. Minor changes may be caused by the inclusion of new items or the disappearance of old ones, or by periodical revision of the weighting system.

When a major change is made it is customary to begin again with a new base, frequently the year on which the new weighting system is based. In this case it is not usually necessary to recalculate the old index number on the same system. It is sufficient to splice or blend the two series as follows—

Suppose the old series is—

1963—100
1964—108
1965—111
1966—120
1967—125

and the new series is—

1967—100
1968—114
1969—121
1970—124

The whole of the first series is multiplied by 0·8, the reciprocal of 1·25, so that the index for 1967 is 100 in both series. The combined series is then—

1963— 80
1964— 86
1965— 89
1966— 96

1967—100
1968—114
1969—121
1970—124

Methods of Calculation

Broadly speaking, methods of calculating an index fall into two categories, one based on *relatives* and the other on *aggregates*. In certain circumstances, as will be shown later, the two methods are equivalent.

A price relative is simply the current price of a commodity divided by the base price. Like the index itself, it is generally expressed as a percentage. Thus, if the price of an article is £15 and its average price in the base year was £12, the price relative will be—

$$\frac{15}{12} \times 100 = 125$$

Relatives of output or other quantities are defined similarly. An obvious method of compiling an index, therefore, is to calculate relatives for all constituent items and average them in some way. The next few sections will describe several ways of doing this.

The other method is to consider an aggregate or "basket" of goods, e.g. the goods actually bought or sold in a given year, and compare their value at current prices with their value at base-year prices. This method will also be discussed in some detail.

Naturally, these different methods will give slightly different results. This is not a serious objection to index numbers as such, for unless price movements are very erratic or conditions are changing so greatly that comparisons mean little in any case, it will be found that the differences are not serious. Some methods are better and some indices more accurate than others, but it is seldom possible to say that one is right and the others wrong. An index number is like a slide rule; its purpose is to give a fair estimate, and this it generally does.

The Arithmetic Mean of Relatives

One of the simplest ways of constructing an index number is to take the arithmetic mean of the relatives of the various items, as shown in Example 16.1.

Example 16.1

Calculate a price index for 1970 based on 1965 as 100 for a group of three commodities whose average prices in 1965 and 1970 were as follows—

Commodity	1965 Price	1970 Price
A	£40 per ton	£50 per ton
B	£5 per cwt	£7 per cwt
C	£50 per gross	£58 per gross

It is easily seen that the price relatives for A, B and C are 125, 140, and 116 respectively. Their simple arithmetic mean is—

$$\tfrac{1}{3}(125 + 140 + 116) = 127$$

In a simple AM it is implied that all items are equally important, but if (as generally happens) they are not, the relatives can be weighted by appropriate factors. Suppose, therefore, that it had been decided to give Commodities A, B and C weights of 6, 3 and 1 respectively; then the index for 1970 (1965 = 100) would have been—

$$\frac{125 \times 6 + 140 \times 3 + 116 \times 1}{6 + 3 + 1} = 128\cdot6$$

In general, if there are n commodities with base-year prices $p_0', p_0'', \ldots p_0^n$ and current prices $p_1', p_1'', \ldots p_1^n$, the simple or unweighted AM is (omitting the factor 100)—

$$\frac{1}{n}\left\{\frac{p_1'}{p_0'} + \frac{p_1''}{p_0''} + \ldots + \frac{p_1^n}{p^n}\right\} = \frac{1}{n}\Sigma\left(\frac{p_1}{p_0}\right) \qquad (16.1)$$

and the weighted AM with weights w', w'', $\ldots w^n$ is

$$\frac{1}{\Sigma w}\left\{w'\frac{p_1'}{p_0'} + w''\frac{p_1''}{p_0''} + \ldots + w^n\frac{p_1^n}{p_0^n}\right\} = \frac{1}{\Sigma w}\Sigma\left(w\frac{p_1}{p_0}\right) \qquad (16.2)$$

In these expressions the superscripts denote the several commodities and the subscripts the years concerned, e.g. p_0^n denotes the price of the nth commodity in the base year, not a base-year price raised to the nth power.

A disadvantage of the AM with fixed weights is that it gives too much weight to items that have risen in price, i.e. there is an upward bias in it. When a commodity has suffered a great increase in price, it is generally because it has become scarce, and when it becomes cheap it is generally because it is plentiful. One solution is to vary the weighting, but this method also has its drawbacks, and it may be difficult to decide what weights to use.

The Geometric Mean

The geometric mean of n price relatives is simply the nth root of their product (see page 98), and it is easily found with the aid of logarithms. The formula for the simple or unweighted GM is—

$$P = \sqrt[n]{\left(\frac{p_1'}{p_0'} \times \frac{p_1''}{p_0''} \times \dots \times \frac{p_1^n}{p_0^n}\right)} \qquad (16.3)$$

or, taking logarithms,

$$\log P = \frac{1}{n} \Sigma \log \frac{p_1}{p_0} \qquad (16.4)$$

Example 16.2
Calculate a price index for 1970 (1965 $=$ 100) for the data of Example 16.1, using the geometric mean.

Commodity	Price relative	Log	Anti-log
A	125	2·0969	
B	140	2·1461	
C	116	2·0645	
Total		3)6·3075	
		2·1025	126·6

The answer, 126·6, is slightly less than the AM, 127·0, as it must be, but the difference is not very material. Indeed, it might be argued that the AM and GM would only differ appreciably if the price relatives varied so widely that a price index had little meaning.

It is not really necessary to calculate each price relative with the GM as it is with the AM, since formulae (16.3) and (16.4) may be expressed as—

$$P = \sqrt[n]{\left(\frac{p_1' p_1'' \dots p_1^n}{p_0' p_0'' \dots p_0^n}\right)} \qquad (16.3a)$$

and

$$\log P = \frac{1}{n} \left(\Sigma \log p_1 - \Sigma \log p_0\right) \qquad (16.4a)$$

In the GM, as in the AM, items can be weighted according to their importance, although the weights are no longer factors but indices in the usual algebraic sense of the word. The weighted

GM of n items with weights w', w'', ... w^n is given by—

$$P = {}^{(\Sigma w)}\!\!\sqrt{\left[\left(\frac{p_1'}{p_0'}\right)^{w'}\left(\frac{p_1''}{p_0''}\right)^{w''} \cdots \left(\frac{p_1{}^n}{p_0{}^n}\right)^{w^n}\right]} \qquad (16.5)$$

or

$$\log P = \frac{1}{\Sigma w}\, \Sigma\left(w \log \frac{p_1}{p_0}\right) \qquad (16.6)$$

The principle is really the same in both cases. An item with a weight of, say, 15, is treated as if it were 15 separate items.

Obviously the GM is more troublesome to compute than the AM and not so easy to interpret, but it has certain theoretical advantages over the AM. Before considering these further, however, it will be convenient to deal with the method of aggregates.

The Aggregative Method

This is the best method to use if the data are available. It consists, as already mentioned, in taking a "bundle" of goods and comparing their value at current prices with their value at base prices. The bundle is generally one of two things:

(i) The quantities q_0', q_0'', ... $q_0{}^n$ actually bought or sold in the base year, or

(ii) The quantities q_1', q_1'', ... $q_1{}^n$ actually bought or sold in the current year (or month, etc.).

In the first case the index measures the change in cost of a fixed bundle of goods. The formula, named after Laspeyres, is—

$$P = \frac{q_0'p_1' + q_0''p_1'' + \ldots + q_0{}^np_1{}^n}{q_0'p_0' + q_0''p_0'' + \ldots + q_0{}^np_0{}^n} = \frac{\Sigma q_0 p_1}{\Sigma q_0 p_0} \qquad (16.7)$$

$$= \frac{\text{base-year quantities at current prices}}{\text{base-year quantities at base-year prices}}$$

In the second case the bundle varies from year to year or from month to month, and it is possible for the index to change even when all prices are unchanged. The formula, named after Paasche, is as follows

$$P = \frac{q_1'p_1' + q_1''p_1'' + \ldots + q_1{}^np_1{}^n}{q_1'p_0' + q_1''p_0'' + \ldots + q_1{}^np_0{}^n} = \frac{\Sigma q_1 p_1}{\Sigma q_1 p_0} \qquad (16.8)$$

$$= \frac{\text{current quantities at current prices}}{\text{current quantities at base-year prices}}$$

It does not matter whether all the prices are in the same units as long as the pq products are. Nor, of course, does it matter whether we write $\Sigma q_0 p_1$ or $\Sigma p_1 q_0$.

Example 16.3

Calculate Laspeyres and Paasche price indices for 1970 (1963 = 100) from the following sales data.

Commodity	Sales in 1963		Sales in 1970	
	Tons	£	Tons	£
A	200	6,000	500	20,000
B	420	10,500	400	14,000
C	2,000	20,000	1,500	30,000
D	5	3,500	45	36,000
Total		40,000		100,000

To do this we must calculate the average prices for each year and compile a second table.

Commodity	p_{63}	q_{70}	$p_{63}q_{70}$	p_{70}	q_{63}	$p_{70}q_{63}$
A	30	500	15,000	40	200	8,000
B	25	400	10,000	35	420	14,700
C	10	1,500	15,000	20	2,000	40,000
D	700	45	31,500	800	5	4,000
Total			71,500			66,700

The totals $\Sigma p_{63}q_{63}$ and $\Sigma p_{70}q_{70}$ are, of course, already calculated.
   The Laspeyres price index is—

$$P_L = \frac{66,700}{40,000} \times 100 = 167$$

and the Paasche price index is—

$$P_P = \frac{100,000}{71,500} \times 100 = 140$$

Calculations for a series of years are simpler with the Laspeyres formula, since the denominator, being the total value for the base year, is constant and need only be calculated once.

These aggregative formulae are equivalent to the weighted AM of price relatives, with suitably chosen weights. Thus the Laspeyres formula can be written—

$$P = \frac{p_0'q_0'\left(\dfrac{p_1'}{p_0'}\right) + p_0''q_0''\left(\dfrac{p_1''}{p_0''}\right) + \ldots + p_0^nq_0^n\left(\dfrac{p_1^n}{p_0^n}\right)}{p_0'q_0' + p_0''q_0'' + \ldots + p_0^nq_0^n} \qquad (16.7a)$$

which is the weighted AM of p_1'/p_0', etc. with weights $p_0'q_0'$, $p_0''q_0''$, etc. These products are the actual values of sales of the various items in the base year. Similarly, the Paasche Formula is the AM of price relatives with weights $p_0'q_1'$, etc., the values of current sales at base-year prices.

Index Numbers of Quantity

Indices of quantity or volume present their own problems. There is no difficulty with a fairly homogeneous product like coal or steel, but what is meant, for instance, by the volume of production of the chemical industry? Obviously not simply the physical volume of the many items involved. To add together heavy chemicals, dyes, plastics and drugs and express the total in tons or gallons would be ludicrous. Somehow these things must be weighted according to their importance. A ton of penicillin must carry far more weight in the statistical sense than a ton of soda ash.

The method usually employed is to calculate quantity relatives for each of the commodities included in the index, and take their weighted arithmetic mean, the weights being roughly proportional to the value of production, sales or whatever may be more appropriate, in the base year. Suppose, for example, that we require an index of sales volume. Then with the same notation as before, and denoting the volume index by Q,

$$Q = \frac{1}{\Sigma w}\left\{w'\frac{q_1'}{q_0'} + w''\frac{q_1''}{q_0''} + \ldots + w^n\frac{q_1^n}{q_0^n}\right\} = \frac{1}{\Sigma w}\Sigma\left(w\frac{q_1}{q_0}\right)$$

$$(16.9)$$

If the weights are exactly equal to the base year values, so that $w' = q_0'p_0'$, etc., formula (16.9) becomes

$$Q = \frac{q_1'p_0' + q_1''p_0'' + \ldots + q_1^np_0^n}{q_0'p_0' + q_0''p_0'' + \ldots + q_0^np_0^n} = \frac{\Sigma q_1 p_0}{\Sigma q_0 p_0} \qquad (16.10)$$

We have thus obtained our index as the ratio of two total values, one of current quantities, the other of base year quantities, both valued at base year prices.

It may be objected that the purpose of the index is to measure physical volume, not value. The answer to that is (i) that a multitude of dissimilar things cannot be added together in any sensible unit except a unit of money (after all, money was devised for the very purpose of providing a common unit), (ii) that the main objection to an index based on value is that the value of money itself varies, and this objection does not apply to an index of value at fixed prices.

An alternative but generally less meaningful index of volume would be the ratio of total values at current prices, i.e.

$$Q = \frac{\Sigma q_1 p_1}{\Sigma q_0 p_1} \qquad (16.11)$$

Formulae (16.10) and (16.11) might be called the Laspeyres and Paasche quantity indices respectively. If we denote them by Q_L and Q_P, and similarly for the price indices, we see that

$$P_L \times Q_P = \frac{\Sigma p_1 q_0}{\Sigma p_0 q_0} \times \frac{\Sigma q_1 p_1}{\Sigma q_0 p_1} = \frac{\Sigma p_1 q_1}{\Sigma p_0 q_0}$$

and that

$$P_P \times Q_L = \frac{\Sigma p_1 q_1}{\Sigma p_0 q_1} \times \frac{\Sigma q_1 p_0}{\Sigma q_0 p_0} = \frac{\Sigma p_1 q_1}{\Sigma p_0 q_0}$$

so that both products are equal to the index of total value. Our quantity index could, in fact, have been obtained by dividing the index of total value by the price index, or vice-versa. One, however, would be of the Laspeyres type and the other of the Paasche type, and it is unlikely that both would be ideal for their purpose. In other words, we cannot have it both ways.

Example 16.4

Calculate Laspeyres and Paasche indices of sales volume from the data of Example 16.3

$$Q_L = \frac{\Sigma q_{70} p_{63}}{\Sigma q_{63} p_{63}} = \frac{71,500}{40,000} \times 100 = 179$$

and

$$Q_P = \frac{\Sigma q_{70} p_{70}}{\Sigma q_{63} p_{70}} = \frac{100,000}{66,700} \times 100 = 150$$

The reader should verify that he gets the same results by taking the index of total value, which is 250, and dividing by the appropriate price index as found in Example 16.3.

Hybrid Formulae

The Laspeyres Formula, like its equivalent, the weighted AM of price relatives, tends to overstate a rise in prices because it takes no account of the fall in output or demand that often accompanies a marked increase in price, or of the expansion of production that causes a fall in price. An item may be priced right out of the market, but it retains the weight appropriate to a normal price. For similar reasons, the Paasche index goes to the other extreme and tends to understate the rise in prices.

The two formulae may be regarded as giving upper and lower limits, the true index (if there is such a thing) being somewhere between them. Various compromises have been suggested. One is to average or add together the base-year and current-year quantities, giving the formula—

$$P = \frac{\Sigma p_1(q_0 + q_1)}{\Sigma p_0(q_0 + q_1)} \qquad (16.12)$$

Another is to take the AM or the GM of the Laspeyres and Paasche indices. In practice the difference between the AM and the GM is negligible, but the GM is preferred because, as will be shown later, it satisfies certain mathematical tests. The formula is—

$$P = \sqrt{\left(\frac{\Sigma q_0 p_1}{\Sigma q_0 p_0} \times \frac{\Sigma q_1 p_1}{\Sigma q_1 p_0}\right)} \qquad (16.13)$$

This is called Fisher's "Ideal" Formula, after its chief exponent, the late Professor Irving Fisher (not to be confused with the British statistician, Professor Sir Ronald Fisher).

The principal formulae for calculating index numbers have now been described, viz. the AM and the GM of relatives, simple or weighted, the Laspeyres and Paasche aggregative formulae, and combinations or variations of them. Before considering their relative merits in detail, it will be necessary to describe the Chain-base Method and the theoretical tests mentioned above.

The Chain-base Method

Up to now it has been assumed that each year is compared directly and independently with the base year. Sometimes, however, it is preferable to compare each year with the previous year and build up the resulting "links" into a chain by successive multiplication. This is called the Chain-base Method. Suppose, for example, the index for 1969 based on 1968 is 112·5 and the index for 1970 based on 1969 is 105·6; then the index for 1970 based on 1968 is—

$$112·5 \times \frac{105·6}{100} = 118·8$$

This method provides a more direct comparison between successive years than the Fixed-base Method and makes it easy to change the base year when desired. Its chief advantage is that it takes account of the changing pattern of sales, etc., and in particular it enables new products to be introduced into the index as soon as there are two years' prices to compare. It also enables old products to be retained as long as possible. With the Fixed-base Method, only products common to the current year and the base year could be used. Reclassification of products also creates difficulties, but if both classifications are available for the year of transition, it can be linked up with both the previous year and the following year, giving a continuous index.

A disadvantage of the Chain-base Method is that any defect or abnormality in the index for one year is perpetuated in all subsequent years. The chain is only as strong as its weakest link.

Tests for Consistency

As has been noticed several times in this chapter, certain index-number formulae are not entirely self-consistent, and several theoretical tests have been suggested for judging the suitability of these formulae. Two such tests can be applied to index numbers relating two years with one another; these are—

(i) the time-reversal test,
(ii) the factor-reversal test.

A third test, called the circular test, concerns the relations between three or more years. These three tests will be discussed in turn.

1. The *time-reversal test* implies that a price index, like an individual price, should show the same relative movement from one year to another whichever year is taken as base, i.e. that if P_{12} denotes the index for Year 2 based on Year 1, and P_{21} that for Year 1 based on Year 2,

$$P_{12} : 100 = 100 : P_{21}$$

or, disregarding the factor 100,

$$P_{12}P_{21} = 1 \qquad (16.14)$$

To take an extreme case, which would not occur in practice, suppose there are only two commodities A and B, equally weighted, and that the price of A remains constant, while the price of B in Year 2 is four times its price in Year 1. Then, with Year 1 as base, the relatives are 100 and 400, and with Year 2 as base, they are 100 and 25.

Now with the AM,

$$P_{12} = \tfrac{1}{2}(100 + 400) = 250$$
$$P_{21} = \tfrac{1}{2}(100 + 25) = 62\cdot5$$

whereas with the GM,

$$P_{12} = \sqrt{(100 \times 400)} = 200$$
$$P_{21} = \sqrt{(100 \times 25)} = 50$$

Clearly, the GM is time-reversible and the AM is not.

2. The idea behind the *factor-reversal test* is that a formula involving both prices and quantities, if right for a price index, should be right for a quantity index when p's and q's are interchanged, and that the product of the two indices should be equal to V, the index of value. This does not apply to any formula which contains p's only. If the p's and q's are interchanged in the Laspeyres Formula,

$$P = \frac{\Sigma p_1 q_0}{\Sigma p_0 q_0},$$

$$Q = \frac{\Sigma q_1 p_0}{\Sigma q_0 p_0},$$

and

$$PQ = \frac{(\Sigma q_1 q_0)\,(\Sigma q_1 p_0)}{(\Sigma p_0 q_0)^2},$$

which is not equal to V, so that the test is not satisfied.

Similarly with the Paasche Formula. In fact, the only formula given in this chapter that satisfies the factor-reversal test is the "Ideal" Formula.

From formula (16.13),

$$P = \sqrt{\left(\frac{\Sigma p_1 q_0}{\Sigma p_0 q_0} \times \frac{\Sigma p_1 q_1}{\Sigma p_0 q_1}\right)}$$

and interchanging p's and q's,

$$Q = \sqrt{\left(\frac{\Sigma q_1 p_0}{\Sigma q_0 p_0} \times \frac{\Sigma p_1 q_1}{\Sigma q_0 p_1}\right)}$$

$$PQ = \sqrt{\left(\frac{\Sigma p_1 q_0}{\Sigma p_0 q_0} \times \frac{\Sigma p_1 q_1}{\Sigma p_0 q_1} \times \frac{\Sigma p_0 q_1}{\Sigma p_0 q_0} \times \frac{\Sigma p_1 q_1}{\Sigma p_1 q_0}\right)}$$

$$= \frac{\Sigma p_1 q_1}{\Sigma p_0 q_0} = V,$$

and the test is satisfied.

3. The *circular test* is an extension of the time-reversal test to more than two years. If P_{12}, P_{23}, etc. are indices (omitting the factor 100) for Year 2 based on Year 1, for Year 3 based on Year 2, and

so on, the circular test requires that—

$$P_{12}P_{23} \ldots P_{n-1, n}P_{n1} = 1 \qquad (16.15)$$

When $n = 2$ this equation reduces to formula (16.14). It follows that $P_{1n}P_{n1} = 1$, and consequently that

$$P_{1n} = P_{12}P_{23} \ldots P_{n-1, n}$$

The circular test is thus more stringent than the time-reversal test, as it includes the latter as a special case, and it also implies that under any formula satisfying the test the chain-base and fixed-base methods give the same result.

One formula that satisfies the circular test is the aggregative formula with fixed weights, i.e.

$$P = \frac{\Sigma wp_1}{\Sigma wp_0} \qquad (16.16)$$

For simplicity this will be shown for three years only.

$$P_{12}P_{23}P_{31} = \frac{\Sigma wp_1}{\Sigma wp_0} \times \frac{\Sigma wp_2}{\Sigma wp_1} \times \frac{\Sigma wp_0}{\Sigma wp_2} = 1$$

Another formula that satisfies the circular test is the geometric mean, unweighted or with fixed weights.

Fisher's "Ideal" Formula, for which somewhat extravagant claims have been made by some of its advocates, fails to satisfy the circular test.

What Makes a Good Index Number?

It might be thought, from the amount of space devoted to tests for consistency in this chapter and in other literature on index numbers, that they gave a perfect criterion for a good formula. In the author's opinion, however, there are other things at least equally important. First, an index number should be simple in conception and easily interpreted. The Laspeyres Formula scores heavily here. The man in the street can understand an index that tries to measure the changing cost of the things he bought in a particular year. The Paasche Formula is also easily understood, and so is the AM of price relatives, but the GM is more difficult to interpret. So is the "Ideal" Formula, except as a compromise between two formulae that are biased in opposite directions.

Secondly, an index should be reliable—not necessarily precise, but reasonably accurate for its purpose. An unweighted AM or an AM with fixed weights is liable to give misleading results—generally too high—as time goes on and the weighting becomes obsolete. The GM is an attempt to correct the upward bias of the AM, and in practice the GM with base-year weights is fairly accurate for small movements, but an unweighted GM or AM is a confession of ignorance, for it suggests that the compiler has no idea what the correct weighting is.

Thirdly, an index number should be reasonably consistent. This is really a corollary of the second point, for if it is grossly inconsistent it cannot be accurate. For example, the effect of changing the base year, and revising the weighting at the same time, should be to make slight, not drastic, changes in relative movements of the index.

The author does not take the view that an index that satisfies the various mathematical tests is therefore the right index. Indeed, there is no such thing as a single "right index," appropriate in all circumstances. The Sales Manager may say to the statistician, "Our sales have doubled in value in the last ten years. How much of that is increase in prices and how much is increase in volume?" As Examples 16.3 and 16.4 have shown, there is no simple and unique answer.

High claims have been made by many people for Fisher's "Ideal" Formula, but it is rarely used because—

(a) it involves twice as much work as the Laspeyres or Paasche Formula separately,

(b) it is not quite so easy to interpret,

(c) current quantities (or in the case of an index of quantity, current prices) are often unknown, and

(d) even the "Ideal" Formula is not consistent for a series of years, but only between the old base year and the new. In fact, if conditions are constantly changing, necessitating frequent revision of the weighting, it is impossible to satisfy all tests of consistency at the same time.

This raises an important distinction between long-term and short-term movements. In a monthly series covering a few years only, especially if there are marked seasonal movements, fixed weights are generally best, for with current weighting it would be possible for variations in the pattern of sales, imports, etc. to cause fluctuations in the index even when prices remained constant. Fixed weights ensure that changes in the index really do reflect changes in prices. Over a long period, however, changes in pattern must be allowed for. This can be done either by using current weighting or by running a fixed weight index for a few years, then revising the weighting and linking up the two series.

Exercises

16.1 Calculate a price index for fresh fruit in 1960 and 1962 using 1958 as the base year.

	Base year 1958 weights	Base year 1958 prices pence	1960 prices pence	1962 prices pence
Apples	320	4	5	3
Oranges	30	3	$3\frac{1}{2}$	$4\frac{1}{2}$
Lemons	40	8	9	7
Pears	110	4	$4\frac{1}{2}$	4

What purpose is served by using weights in the construction of a price index? (ACCA)

16.2 From the following information compute the "All items" index numbers of retail prices using as the weights (a) those of the UK Index of Retail Prices and (b) 1955 expenditure of all consumers. Comment on your results.

UK Index of retail prices (17 January 1956 = 100)			
Group	Weights	Group index (October 1956)	Personal expenditure*
Food	350	101·8	4,136
Drink and tobacco	151	104·0	1,739
Clothing	106	101·0	1,268
Housing	87	104·5	1,065
Fuel and light	55	102·4	521
Household goods	66	101·3	903
Other goods and services	185	104·0	3,151
All items	1,000		12,783

* £000,000—All consumers, at current market prices.

(IoT)

16.3 From the following data, calculate an index number for measuring the change in the general level of prices of the group of goods listed.

Commodity	Unit of purchase	Prices		Quantities	
		Period 0 (p_0) pence	Period 1 (p_1) pence	Period 0 (q_0)	Period 1 (q_1)
Bread	2 lb	18	21	100	90
Tea	1 lb	80	90	10	8
Potatoes	7 lb	49	35	20	30
Butter	1 lb	56	56	30	30
Milk	Quart	21	18	7	10
Beef	1 lb	42	63	80	60
Mutton	1 lb	30	40	70	60

Give reasons for the method of index number calculation chosen.

(ACCA)

16.4 From the following information calculate the *percentage* change between 16 October 1962 and 15 October 1963 in the retail prices of—

(*a*) transport and vehicles,

(*b*) all items except transport and vehicles,

(*c*) all items including transport and vehicles.

Group	Percentage weight	Price index	
UK Index of retail prices (16 January 1962 = 100)			
		16 Oct 1962	15 Oct 1963
Food	32	100·5	104·2
Alcoholic drink	6	100·6	103·2
Tobacco	8	100·0	100·0
Housing	11	104·9	109·8
Fuel and light	6	101·1	104·9
Durable household goods	6	100·8	100·3
Clothing and footwear	10	103·0	103·7
Transport and vehicles	9	101·1	100·5
Miscellaneous goods	6	101·1	102·6
Services	6	102·9	104·9
	100		

(IoT)

16.5 The table below shows the prices and the annual consumption of the major raw materials used in a particular brewery in 1958 and 1966—

	1958		1966	
Material	Price per ton £	Con- sumption tons	Price per ton £	Con- sumption tons
Malt	49	19,874	46	25,116
Hops	512	732	724	496
Sugar	46	1,865	51	2,486
Wheat flour	31	873	27	2,093

Calculate a current-weighted index number showing the overall change in raw material prices.

Why is it very unlikely that you would have obtained the same result if you had used a base-weighted index? (ICWA)

16.6 The following figures relate to all the exports of a certain country over a period of 10 years—

	Value of exports in £ million based on—		
Year	(1) Actual prices and quantities exported each year	(2) Actual quantities exported but prices in year 1	(3) Actual prices but quantities exported in year 1
1	100	100	100
2	107	98	110
3	112	97	119
4	115	94	132
5	116	91	140
6	115	90	150
7	112	89	162
8	107	86	171
9	100	83	179
10	91	82	191

(a) Column 2 provides an index for quantities exported each year, using year 1 as base. Calculate an alternative index for quantities, using year 1 as base.

(b) Compare your own index with that given in the table above, explaining why any differences may have arisen. State, with reasons, which type of index number you prefer here. (IoS)

16.7 Production and prices of a country's three main vegetables in the years 1962, 1963 and 1964 are given below—

Vegetable	1962		1963		1964	
	Thousand tons	Price/ton £	Thousand tons	Price/ton £	Thousand tons	Price/ton £
Potatoes	5,000	3	6,000	2	7,000	3
Cabbages	4,000	6	4,000	6	5,000	5
Turnips	8,000	4	9,000	6	12,000	7

Construct two different indices of physical volume. Quote the formulae used in their construction and state briefly the differences in meaning and use between the two indices.   (IoS)

16.8 The following data refer to exports of fertiliser in 1958 and 1968

	Quantity ('000 tons)		Values (£'000)	
	1968	1958	1968	1958
Nitrogenous	63·1	24·2	2,672	389
Other	736·3	459·1	7,929	1,833
Totals	799·4	483·3	10,601	2,222

(a) Find the average value for each type of export in each year. Revalue 1968 quantities at 1958 values and 1958 quantities at 1968 values.

(b) Calculate index numbers of average values and of volume

(i) for 1968 on 1958 as base,

(ii) for 1958 on 1968 as base.   (IoS)

17

Introduction to Published Statistics

THE remaining chapters of this book deal almost entirely with the principal sources, methods of compilation, and interpretation of published statistics, i.e. data published by official bodies or private bodies with a recognized status. Not only will this give the student a working knowledge of the statistics themselves, but it will illustrate many of the principles expounded in earlier chapters. No attempt is made to cover the whole field, and many topics such as finance, education and crime are omitted. As a general rule, only British or international statistics are considered.

The reader must remember that this book is going to the printer in 1971, and that, although every effort has been made to give up-to-date information, by the time he reads these chapters there will almost certainly be changes in the scope and content of some of the statistics described. He should therefore endeavour to keep them up to date for himself.

Before considering the publications themselves, it is desirable to give a brief account of the background and development of official statistics in the United Kingdom.

The Development of British Official Statistics

Like Topsy in *Uncle Tom's Cabin*, official statistics "just growed," slowly and painfully, as a by-product of administrative requirements. Before the nineteenth century they were virtually non-existent, and estimates of population, production, etc. were extremely uncertain. At one point in our history it was commonly thought that the sex ratio was about 2 : 1, i.e., two females to one male. However, a step towards reliable demographic statistics was taken in 1801 with the first Census of Population. But more about that in Chapter 18.

Statistics of the country's economy were much later developing. There was no Census of Production until 1907. The author once had to compile figures of UK production of sulphuric acid, going back as far as possible. For the early 1800's there were only scattered and very crude estimates for isolated years, based on the output of individual firms. From about 1850 onward it was possible to derive rough estimates from imports (taken as equal to consumption) of pyrites. Not until the present century were reliable figures of sulphuric acid production forthcoming.

The inter-war years saw considerable progress both in statistical theory and its applications and in the development of economic statistics, together with a growing awareness of the inadequacy of the available data. The Second World War highlighted the need for reliable figures and for better organization, and in 1941 the Central Statistical Office (CSO for short) was set up to co-ordinate the statistical work of the various Government Departments. For instance, up till then the Board of Trade, the Ministry of Labour and other Government Departments had used different classifications of industry, and it was difficult to relate, say, the figures of employment and unemployment to the statistics given in the Census of Population or the Census of Production. In 1948, therefore, the CSO issued a Standard Industrial Classification in order to secure uniformity and comparability in the statistics of the various Departments. It is now used for the Censuses of Population, Production and Distribution, and other official statistics.

Since the war, published statistics have grown rapidly, but never quite fast enough. In 1956, Harold Macmillan, then Chancellor of the Exchequer, made his famous speech complaining of the deficiencies in the available statistics and the delay in publishing them ("We are always, as it were, looking up a train in last year's Bradshaw"). After that, there was a considerable expansion in statistics of national income, balance of payments, production, distribution, stocks, capital expenditure, etc., particularly of quarterly figures. For production and distribution, in particular, there developed a three-tier structure comprising a Census of Production or Distribution roughly every five years, less detailed annual surveys between censuses, and very detailed monthly or quarterly statistics obtained from a sample of firms as indicators of short-term movements. (See Chapters 21 and 22.) More recently we have had much more adequate statistics of finance and housing, and many figures are now compiled on a regional basis.

The latest development is the formation of a Business Statistics Office, set up in 1969 not only to further the collection and integration of business statistics but also to provide a valuable service to business. One of its main tasks is to compile a Central Register of

Businesses, which will replace the separate registers now held by different Government Departments, provide a frame for sample inquiries and show the interrelations between different reporting units. At the time of writing the situation is still very fluid.

British Statistical Publications

Most Government Departments have their own Statistical Divisions or Sections, and some publish their own journals, the best known being the weekly *Trade and Industry* (formerly the *Board of Trade Journal*), and the monthly *Department of Employment Gazette*. Many publish annual reports, such as the *Annual Report of Inland Revenue*, the *Annual Report of Customs & Excise*, and others which will be mentioned in later chapters. The most important statistical publications, however, are compiled by the Central Statistical Office, the chief of them being—

1. *The Monthly Digest of Statistics*, which, as its name indicates, is an official summary of statistics published each month, containing about 170 tables. The main headings are—

I	National income and expenditure
II	Population and vital statistics
III	Labour
IV	Social services
V	Agriculture and food
VI	Production, output and costs
VII	Energy
VIII	Chemicals
IX	Metals, engineering and vehicles
X	Textiles and other manufactures
XI	Construction
XII	Retailing and catering
XIII	Transport
XIV	External trade
XV	Overseas finance
XVI	Home finance
XVII	Wages and prices
XVIII	Entertainment
XIX	Weather

An annual Supplement, entitled *Definitions and Explanatory Notes*, provides a useful account of the tables shown and units employed in the *Monthly Digest*.

2. *Economic Trends*, a monthly booklet of small charts and tables of key statistics designed to show the principal trends in the economic life of the United Kingdom. It is largely a condensed

pictorial version of the *Monthly Digest*, but there are also special articles in each issue dealing with specific aspects of economic statistics.

3. The *Annual Abstract of Statistics*, covering much the same ground as the *Monthly Digest* and a good deal more, containing about 400 tables. With a few exceptions it gives only annual figures and usually covers a period of at least ten years.

4. *National Income and Expenditure of the United Kingdom*, an annual Blue Book which will be discussed at length in Chapter 24.

5. *Statistical News*, a quarterly booklet first published in 1968 as part of the new service to industry and the public, to keep them informed about current developments.

The CSO also publishes a series of *Studies in Official Statistics* on subjects such as the Index of Industrial Production and the Index of Retail Prices. These should not be confused with the excellent *Guides to Official Sources* prepared by the Interdepartmental Committee on Social and Economic Research, of which there are six, i.e.,

1. Labour Statistics.
2. Census Reports of Great Britain, 1801–1931.
3. Local Government Statistics.
4. Agriculture and Food Statistics.
5. Social Security Statistics.
6. Census of Production Reports.

These *Guides* are most valuable as books of reference.

A considerable number of statistical yearbooks and bulletins are published by trade associations and other bodies, e.g.—

World Metal Statistics
Iron and Steel Industry Monthly Statistics
The Motor Industry of Great Britain (annual)
Rubber Statistical Bulletin (monthly)
Lloyd's Shipping Register (quarterly)

Statistical series and articles are also published by a number of unofficial or semi-official bodies, notably in the *National Institute Economic Review*, published quarterly by the National Institute of Economic and Social Research.

International Statistics

The Statistical Office of the United Nations produces a large number of publications, frequently in both English and French and sometimes in Spanish also. The *Monthly Bulletin of Statistics* gives fairly up-to-date figures of employment, production, etc., for the various countries of the world. Annual publications are the

Statistical Yearbook, the *Demographic Yearbook* and the *Yearbook of International Trade Statistics*. A very useful series is *Studies in Methods*, dealing with such matters as Index Numbers of Industrial Production, National Accounts, and Industrial Censuses.

The Food and Agriculture Organization also produces a number of yearbooks and a *Monthly Bulletin of Agricultural Economics and Statistics*. The International Labour Office publishes a *Yearbook of Labour Statistics* and a *Monthly Labour Review*.

Another body which publishes international statistics is the OECD (Organization for Economic Co-operation and Development). A *Foreign Trade Statistical Bulletin*, issued in three series, deals with various aspects of the foreign trade of member countries of the OECD. *Main Economic Indicators*, published monthly, contains comprehensive data for the same countries.

Both the United Nations and the OECD have done much to standardize the collection and presentation of statistics in member countries. In particular, the United Nations have brought out a Standard International Trade Classification which has made it much easier to compare the foreign trade statistics of different countries. The trade classification of the United Kingdom has been based on the SITC since the beginning of 1954.

Suggestions for further reading

N.B.—In this and subsequent chapters, references and suggestions for further reading are published by HMSO unless otherwise specified.

DEVONS, E., *An Introduction to British Economic Statistics* (Cambridge University Press)

HARVEY, J. M., *Sources of Statistics* (Clive Bingley)

LEWES, F. M. M., *Statistics of the British Economy* (Allen & Unwin)

BERMAN, L. S., "Recent Improvements in Official Economic Statistics," *J. Roy. Stat. Soc.*, Series A, Vol. 134, Part 4 (1971)

FESSEY, M. C., "Some Developments in Economic Statistics since 1934," *The Statistician*, Vol 19, No 2 (1969)

Studies in Official Statistics, Nos. 5, 9, 10, 12 and 15, *New Contributions to Economic Statistics*. (Articles reprinted from *Economic Trends*.)

Statistical News

Social Trends

18

Demographic Statistics

Demography is the study of people from a statistical point of view, and deals with such matters as population, births and deaths, marriages and fertility. It will also be taken here to include sickness, or, as it is sometimes called, morbidity, but employment and income statistics will be left to later chapters.

Accurate demographic statistics require two things; a periodic census, or stocktaking of the whole population at a given time, and a system of compulsory registration of births, deaths and marriages. In England and Wales the Director and Registrar General, as head of the Office of Population Censuses and Surveys (incorporating the General Register Office), is responsible for these functions. Scotland has its own Registrar-General and General Register Office and publishes separate reports, but the system is very similar to that of England and Wales and there is close co-ordination between the two.

The General Register Office at Somerset House and what is substantially the present system of registration were set up by an Act of 1836. Registration was compulsory from the outset, although penalties for failure to register were not imposed until 1874. In 1874 also, the issue of medical certificates of cause of death was made compulsory, which greatly facilitated the analysis of deaths by cause.

The Census of Population

Necessary as registration is, the basis of all population estimates is the census. The first census in Great Britain was taken in 1801, and from then until 1931 inclusive, a census was taken every tenth year, generally with slight differences between the census for England and

Wales and that for Scotland. Owing to the war, there was no census for 1941, although the National Register taken in September 1939 helped to fill the gap. The decennial census was resumed in 1951, and is being continued, but a sample census was also held in 1966. The post-war censuses are described in some detail below.

The population recorded is *de facto*, not *de jure*. In plain English, this means that a man is recorded for the district in which he happens to be at the time, not necessarily that in which he normally lives. It is therefore desirable to avoid holiday seasons and other times when many people are away from home. The time usually chosen is midnight on a Sunday in April. The population of the country as a whole is also *de facto*; thus it excludes people abroad on business or on holiday and British soldiers serving overseas, but includes American service personnel stationed in England and other foreign visitors.

A Preliminary Report on the Census is usually published within a few months, giving the total population by sex, but not by age, for countries and districts. This is followed some time later by County Volumes giving minute details of the local populations, then by reports on Occupations, Industries, Housing, etc., and finally by the General Report.

One problem still to be overcome is the inevitable delay in converting millions of schedules into statistical tables and publishing the results. The use of high-speed tabulating machines in the 1911 and later Censuses did not greatly reduce this delay, as it stimulated the demand for more detailed tabulations, and not only do the schedules have to be examined and coded as before, but fifty million cards have to be punched by hand before they can be sorted and tabulated. Much of the editing, tabulation and analysis is now done by computer, but the preliminary work still has to be done by hand. However, it is now possible to obtain from the census office (for a small fee) special tabulations or computer tape derived from census data, long before the reports are published.

Every effort is made to ensure accuracy in the census, but some errors are unavoidable. They are chiefly errors of age, marital status and occupation. Uncertainty will cause a man to enter his age as 70 rather than 69; there is a distinct bias towards ages ending in 0, and a smaller bias towards ages ending in 8. Vanity may induce a woman of 45 to give her age as 40, a divorced woman to describe herself as widow, and a clerk to give his occupation as "official" or "executive." Although the standard of enumeration for the population in general is high, there is a tendency to understate the population of certain groups who present special difficulties. This is especially true of the mobile element, for example, vagrants and people living in caravans. Householders often forget to include young babies on the returns. Improvements in the organization of the census, however, are steadily reducing the number of errors made.

The 1951 Census, being the first for twenty years, was more detailed than usual. The information required was as follows—

A	Name and surname
B	Relationship to head of household
C	Usual residence
D	Sex
E	Age
F–K	Particulars as to marriage
L	Birthplace
M	Nationality
N–O	Education
P	Personal occupation
R	Employer and employer's business
S	Place of work
T	Household arrangements for water supply, cooking, etc.

Under Particulars as to marriage, married women under fifty years of age had to state their date of marriage, the number of children born alive in marriage, and whether they had given birth to a live-born child during the past twelve months. There were also questions on language for Wales and Scotland. In order to obtain some preliminary results as quickly as possible, a 1 per cent sample of all schedules was taken and analysed, and the results of that sample were published the following year.

In the 1961 Census the above questions were repeated, with some modifications, but the load on the public was reduced by means of a 10 per cent sample. Nine out of ten householders had only to answer the basic questions A to M, which included a new question on housing tenure. The tenth received a form similar to that of 1951 but containing new questions about—

(a) Change of usual address during the last year, i.e. since 23 April 1960; or, if unchanged, duration of stay at present residence

(b) Qualifications in science or technology

(c) Details of members of the household who were absent on the Census night.

The 1961 Census was also noteworthy for the first "post-enumeration survey" by skilled field staff to check errors of enumeration and response.

The 1966 Census was the first census to be held five years after the previous one. It was also the first to be based entirely on a sample. A sample of schedules had been used for special treatment in 1951, but every individual had been covered; similarly, in 1961 every person was enumerated, but only 10 per cent were given the full range of questions. But the 1966 Census was taken on a random 10 per cent sample of households. One of the main objects was to obtain information about housing, so there were questions about number of rooms, whether the household had a hot water supply, bath, W.C., etc., number of cars owned and where they were kept. There was also a new question about mode of transport to work.

At the time of printing, the Office of Population Censuses and Surveys is preparing the results of the Census held in April 1971. About 100,000 enumerators were required to deliver and later to collect the forms. There were special forms for hotels and boarding houses, ships, etc., but the principal form was the "H Form for Private Households." There were three parts to the H Form, which may be summarized as follows—

Part A contained questions about the household's accommodation—type of tenure, number of rooms, number of cars and vans, and use of baths, toilets, cookers, etc.

Part B asked for details of all persons present on Census night (25 April) or who joined the household on Monday, 26 April and had not been enumerated elsewhere, viz. sex, date of birth, marital status, country of birth (and country of birth of each person's father and mother), whether employed or retired, etc., usual address one year ago and five years ago, educational qualifications, occupation, and means of transport to work. There were also questions about date of marriage and dates of birth of children for women under 60.

Part C asked for brief particulars of any person who usually live in the household but who was not present and for whom, therefore, no entry had been made in Part B. The information so collected will make it possible to present figures of the normal composition of private households, as needed particularly by those engaged in social research.

Birth Rates and Fertility

The *birth rate* for a given year is obtained by dividing the number of live births by the average population during the year and multiplying by 1,000. In practice the estimated population at the end of June is taken as the average for the year. For a shorter period, such as a quarter, the number of births must first be multiplied by four, or the appropriate factor, to give the equivalent annual rate.

The birth rate, although useful over periods of a few years, is not a good measure of the way the population is replacing itself, because so much depends on its age-distribution. Given normal fertility, a young population will produce a higher birth rate than a population with a high proportion of old people. A more useful measure, if the data are available, would be the *fertility rate*, obtained by relating total live births to the number exposed to risk, i.e. the number of women of reproductive age. This is bound to be somewhat arbitrary, but is generally taken as covering ages 15 to 44 inclusive. It is usual to distinguish legitimate and illegitimate fertility rates, but the former includes many births conceived before marriage.

Reproduction rates are different again, being built up from the age-specific fertility rates, i.e., the rates at each age of the potential mothers. Briefly, the *Gross Reproduction Rate* is the average number of daughters being born to women of reproductive age at current fertility rates. The *Net Reproduction Rate* takes mortality into account, being the average number of daughters being born who may be expected to survive to their mothers' ages and replace them as potential mothers. If the Net Reproduction Rate is unity, the population is replacing itself—or so it was once claimed.

It is now realized that Net Reproduction Rates may be misleading if, in the year in question, marriage and fertility rates were abnormal. In the years immediately following the Second World War there was a sharp rise in the number of births, owing to the large number of marriages that took place and the number of reunions between husband and wife who had not previously had children, or at least had had none for several years. For a reliable measure of replacement it is necessary to study each *cohort* of married women, i.e. those married in a particular year, and to analyse the fertility of each cohort by age at marriage and duration of marriage. From this *cohort analysis*, as it is called, it is possible to construct *General Replacement Rates* which take all these factors into account. Even then they are liable to fluctuate owing to temporary influences. The fact is that, as in economic affairs, no single index will suffice; there is no real substitute for a thorough examination of all the facts.

Death Rates

The *crude death rate* or *mortality rate* for a given year is the number of deaths per thousand, the equivalent annual rates being given for periods other than a year. Death rates can also be calculated for particular occupations, ages, or any other group of people exposed to risk. The death rate for a particular age or age-group is called an *age-specific death rate*. The *infant mortality rate* is the number of deaths of infants under one year old per 1,000 live births. At the beginning of this century it was about 150 per 1,000, but it is now about 20 per 1,000.

Table 18.1 shows extracts from English Life Table No 12, 1960–1962 (Males). A life table is constructed from the age-specific mortality rates experienced by a given population during a given period (in this case by males in England and Wales during the years 1960–1962), and shows what would happen to, say, 100,000 babies born if, in every future year of age, their mortality rates were those of the population observed. In Table 18.1, l_x denotes the number who, on this assumption, would reach age x, and d_x the number of those who would die before reaching age $x + 1$; p_x is the proportion of those l_x who would survive to age $x + 1$, and q_x the proportion who would not. Hence—

$$d_x = l_x - l_{x+1}$$

$$q_x = \frac{d_x}{l_x}$$

$$p_x = 1 - q_x = \frac{l_{x+1}}{l_x}$$

The last column shows the *expectation of life*, denoted by $\overset{\circ}{e}_x$, at any age x. This is the average number of years that people aged x years may be expected to live, at current rates of mortality. Females have lower mortality rates than males, and therefore greater expectation of life, at all ages.

Table 18.1   Extract from English Life Table No 12, 1960–1962 (Males)

Age x	l_x	d_x	p_x	q_x	$\overset{\circ}{e}_x$
0	100,000	2,449	0·97551	0·02449	68·09
1	97,551	153	0·99843	0·00157	68·80
2	97,398	96	0·99901	0·00099	67·90
3	97,302	67	0·99931	0·00069	66·97
4	97,235	60	0·99938	0·00062	66·02
.	.	.	.	.	.
.	.	.	.	.	.
.	.	.	.	.	.
65	68,490	2,499	0·96352	0·03648	11·95
66	65,991	2,625	0·96022	0·03978	11·39
67	63,366	2,745	0·95668	0·04332	10·84
68	60,621	2,856	0·95288	0·04712	10·31
69	57,765	2,959	0·94878	0·05122	9·79
.	.	.	.	.	.
.	.	.	.	.	.

The student will find it profitable to consider the following questions—

1. If a man's expectation of life at 30 is 40 years, why is his expectation at 40 more than 30 years?

2. Why might the expectation of life of a child one year old be greater than that of a new-born baby?

3. Why would you expect the average age at death of judges or bishops to be higher than that of, say, journalists or shopkeepers?

Deaths are extensively analysed by cause, occupation and social class, as in the Registrar General's Supplement on Occupational Mortality. All these factors raise difficulties of classification. Changes over time in death rates from certain diseases, for example, may be due partly to changes in classification and improvements in diagnosis. The classification generally followed now, for both mortality and sickness, is the International Statistical Classification of Diseases, Injuries and Causes of Death, issued by the World Health Organization. Mortality rates for occupations are less reliable than for causes, since it is necessary to rely on the description given by the person registering the death. Moreover, the occupation stated may not be the one that the deceased followed for most of his life, and which may even have caused his death. Thus, a factory worker who has to leave the factory on account of his health and spends the last year or two of his working life as a car-park attendant will probably be recorded as following the latter occupation.

Standardized Rates

It will be obvious from Table 18.1 that the crude death rate, like the birth rate, depends greatly on the age distribution of the population, so it may be misleading to compare the death rates of two towns, countries, periods or occupations that differ in their age composition. In 1968, for example, Harlow, a new industrial town, had a death rate of 4·8 per 1,000, whereas Worthing, a seaside resort full of retired people, had a rate of 25·3 per 1,000.

This difficulty can be largely overcome by the use of *standardized death rates*. The local population is broken up into a reasonable number of age-groups, and a mortality rate calculated for each group. These rates are then applied to a standard population, e.g. the population of the country as a whole. Example 18.1 is purposely over-simplified, but it will illustrate the method.

Example 18.1

Calculate a standardized death rate for a town, given the following table.

The crude death rate is 14 per 1,000. The standardized death rate is—

$$\{(10 \times 4) + (1 \times 12) + (2 \times 12) + (6 \times 14) + (25 \times 6)$$
$$+ (120 \times 2)\} \div 50 = 550 \div 50$$
$$= 11 \text{ per } 1,000$$

In practice, the population would be broken down into much finer age-groups and standardized on sex as well as age. The

Age-group	Town population (thousands)	Deaths per 1,000	Actual deaths	Standard population (millions)
0– 4	5	10	50	4
5–19	10	1	10	12
20–39	15	2	30	12
40–59	20	6	120	14
60–74	6	25	150	6
75 and over	4	120	480	2
Total	60	14	840	50

standardized rate is, in effect, a weighted mean of age-specific rates, with the standard population as the weighting system.

The above method, called the Direct Method, is impracticable if the age-specific rates of the towns or districts are unknown, and we then fall back on the somewhat complicated Indirect Method. We first calculate the *Area Comparability Factor* (ACF for short) by dividing the death rate of the standard population by the *Index Death Rate*, the death rate which the town would have experienced if it had had the age-specific rates of the standard population. The crude death rate of the town is then multiplied by the ACF to give the indirect standardized rate.

Example 18.2

Calculate an indirect standardized death rate for the town of Example 18.1, given that the age-specific rates for the six age-groups in the standatd population are 6, 0·5, 1, 7, 35 and 150 per 1,000 respectively.

The calculations are best set out by means of a table.

Age-group	Town population (thousands)	Standard population (millions)	Rates per 1,000 of standard population	Deaths at standard rates — In town	Deaths at standard rates — In standard population (thousands)
0– 4	5	4	6	30	24
5–19	10	12	0·5	5	6
20–39	15	12	1	15	12
40–59	20	14	7	140	98
60–74	6	6	35	210	210
75 and over	4	2	150	600	300
Total	60	50		1,000	650

$$\text{Death rate of standard population} = \frac{650}{50} = 13 \text{ per } 1,000.$$

$$\text{Death rate of town at standard rates} = \frac{1,000}{60} = 16{\cdot}7 \text{ per } 1,000$$

$$\therefore \text{ACF} = \frac{13}{16{\cdot}7} = 0{\cdot}78$$

From Example 18.1, the crude death rate of the town is 14 per 1,000.

$$\therefore \text{ indirect standardized rate} = 14 \times 0{\cdot}78$$

$$= 10{\cdot}92 \text{ per } 1,000$$

Another summary measure widely used is the *Standardized Mortality Ratio* (SMR for short), although it is generally given as a percentage. It is simply the actual number of deaths in the town or group expressed as a percentage of the "expected" number that it would have had at the age-specific rates of the standard population. With the data of Examples 18.1 and 18.2, the SMR of the town would be—

$$\frac{840}{1,000} \times 100 = 84$$

A low SMR means that the mortality of the town compares favourably with the standard population, allowing for the difference in age (and sex) structure.

All these summary figures, while useful, have their limitations and tell only part of the story.

Migration

This is probably the weakest point in our population statistics, particularly internal migration from one part of the country to another. Some countries have a system of compulsory registration of such movements, but in Great Britain only aliens are compelled to register. Our migration statistics are derived mainly from the International Passenger Survey. This is a sample survey of passengers entering or leaving the country on all the principal sea and air routes. The sampling fraction varies according to the route and the time of year. Those who say they intend to stay in (or out of) the country for at least a year are recorded as migrants. The grossed-up figures are subject to fairly large sampling errors.

There are also, as we have seen, questions in the Census designed to provide information about migration, both internal and external.

Population Projections

Estimates of the population between censuses are made by taking the numbers recorded at the last census and making adjustments for births and deaths registered since that date, and for migration. Owing mainly to the lack of any comprehensive system of recording migration, as noted above, such estimates are by no means precise, but they are accurate enough for most purposes.

For planning purposes it is necessary to have forecasts of population changes in the future. The Office of Population Censuses and Surveys regularly analyses the trends of fertility, mortality and migration, and projects these trends into the future in collaboration with the Government Actuary, who requires such forecasts for the proper conduct of social insurance. These projections are published annually in the Quarterly Return of the Registrar General (4th quarter) and later in the *Annual Abstract of Statistics*.

Sickness

The chief difficulty about sickness statistics is that sickness as such cannot be recorded and measured. One can only record attendances at the hospital or surgery, absence from work or claims for benefit. Again, much depends on the doctor's diagnosis or willingness to sign a medical certificate. Certain infectious diseases, however, have to be notified, and these are well documented.

After the Second World War the General Register Office carried out a number of experiments in the collection of morbidity statistics. These included the establishment of a permanent and continuous register of cancer patients (a measure of prevalence and a means of ascertaining progress in the effectiveness of surgery and radiotherapy), a register of mental patients (for long-term follow-up), an analysis of hospital in-patient records (as a measure of sickness treated in hospital), and an analysis of general-practitioner records (as a measure of sickness treated in general practice). The results of these statistical arrangements have been published either as Supplements to the *Annual Statistical Review* of the Registrar General or as special reports in the series *Studies of Population and Medical Subjects*.

Statistics of claims for sickness and injury benefits and of insured persons absent from work are published by the Department of Health and Social Security. The Department make extensive use of sampling methods in assessing and forecasting the numbers of persons claiming various benefits. Besides maintaining a 2 per cent sample (about 500,000) of the insurance records of the insured population, they take samples (of various sizes) of all family allowances, unemployment benefit claims, retirement pensions, and claims for sickness, injury and disablement benefits.

The subject of demography is vast in itself, and the reader should study some at least of the references given below, particularly the official publications.

Suggestions for further reading

Registrar General's Annual Statistical Review
Registrar General's Quarterly Return of Births, Deaths and Marriages
Registrar General's Weekly Return of Births and Deaths
Reports on the Census of Population of England and Wales
Guides to Official Sources, No 2, "Census Reports of Great Britain 1801–1931"
Annual Report of the Department of Health and Social Security
Digest of Health Statistics for England and Wales
United Nations Demographic Yearbook (UN Statistical Office)
BENJAMIN, B., *Health and Vital Statistics* (Allen & Unwin)
BENJAMIN, B., *Demographic Analysis* (Allen & Unwin)
Cox, P. R., *Demography* (Cambridge University Press)

Exercises

18.1 (*a*) Define a "standardized rate/average" and explain its purpose.
(*b*) For the following data calculate (i) the crude death rate for the Town, and (ii) a standardized death rate.

Age group	Number of deaths	Town population (thousands)	National population (millions)
0– 5	25	2·5	2
6–20	5	5	6
21–40	15	7·5	6
41–60	60	10	7
61–75	75	3	3
76 and over	240	2	1

(*c*) If it is desired to compare the average prices of consumer goods in two countries by means of standardized averages, explain why the choice of standardizing weights is critical to the results. (IoS)

18.2 By means of age-standardized rates, compare the absence (from all causes) of two categories of employees which, for convenience, may be called juniors and seniors, using the data in the following table

Age	Juniors		Seniors	
	Number employed	Percentage lost time	Number employed	Percentage lost time
Under 20	150	2·4	—	—
20–29	350	1·6	20	1·5
30–39	300	2·1	70	3·0
40–49	150	2·8	180	3·5
50–59	50	4·6	240	5·0
60 and over	—	—	90	7·0
Total	1,000	2·2	600	4·5

Comment on your choice of standard population.   (IoS)

18.3 The following table shows particulars of sickness-absence during a recent year for conductors employed by a bus company, compared with what would be "expected" based upon the experience of the conductors employed by all bus operators.

Age group	Average number conductors employed	Actual sickness absence	"Expected" (rate per conductor)
Up to 29 years	50	240 days	6·0 days
30 to 39 years	53	325 days	6·1 days
40 to 49 years	64	515 days	8·1 days
50 to 59 years	59	1,190 days	17·8 days
60 and over	24	630 days	23·2 days
Total	250	2,900 days	

Calculate the overall sickness-absence rates using both the Actual as well as the "Expected" occurrence of sickness. Explain any difference between these two rates in terms of the differences in sickness-absence for the separate age groups.   (IoT)

18.4 Demonstrate the calculation of crude and standardized death rates using the following data—

Age group	Population (in thousands)	Number of deaths	Age distribution of standard population
0– 9	21	350	221
10–24	30	102	298
25–44	37	229	285
45–64	17	354	149
65 and over	5	415	47
Total	110		1,000

(IoS)

18.5 The following table gives the number of deaths of mineworkers in the five years 1949 to 1953 in England and Wales, the 1951 Census population of mineworkers, and the mean annual death rates for all males for the period 1949 to 1953.

Age group	Mineworkers		Annual death rates (all males) per 100,000
	Number of deaths 1949 to 1953	Census population	
20–24	145	16,341	138
25–34	615	67,017	159
35–44	1,255	73,187	287
45–54	3,043	61,718	821
55–64	6,090	35,324	2,295

Calculate the Standardized Mortality Ratio for mineworkers aged 20–64. (IoS)

18.6 An abridged life table for a particular town gives the following estimates of l_x, the number of males surviving to age x out of 1,000 live male births—

Age x years	l_x
0	1,000
1	976
5	971
15	968
45	918
65	685

Each boy starts school on reaching 5 years of age, leaves school on his fifteenth birthday to take a job and retires at age 65. Assuming that the number of male live births in the town is constant at 700 per annum, that mortality is unchanged from that of the above life table and that migration can be ignored, how many school places for boys are needed in the town and how many jobs are required to provide full employment for males? Of those males who start work, what proportion live to draw the retirement pension? (IoS)

19

Prices

For detailed prices it is usually necessary to refer to technical, financial or trade journals. Annual average wholesale prices of many articles over a period of years are given in the *Annual Abstract of Statistics*. They are also given each February in *Trade and Industry*. Prices for many varieties of crops, fruit and vegetables, livestock and other agricultural products are given in *Agricultural Statistics, England and Wales, Part II*, published annually. Most information about prices, however, is published in the form of index numbers, and the rest of this chapter will be devoted to an account of the most important ones used in this country, taking wholesale prices first and retail prices afterwards. Import and export indices will be deferred to Chapter 23.

Wholesale Price Index Numbers

For a detailed account of the old Board of Trade Wholesale Price Index, the reader should refer to earlier editions of this book or to the *Board of Trade Journal* of 24 January 1935. It was designed to measure changes in the value of money relatively to other things, and was derived from 200 price series for all kinds of commodities, including basic materials, intermediate products and manufactured articles. The "All Items" index was the geometric mean of the price relatives of these 200 items.

In 1951 this index was superseded by the present system, which is based on "sectors" of the economy or groups of products. The attempt to compress all price changes into a single index has been abandoned, on the grounds that a family of index numbers which are meaningful to industry is more useful than a single index of somewhat doubtful significance. The geometric mean has also been replaced by the more readily understood arithmetic mean.

There are four tables, published monthly in *Trade and Industry*, with the following titles—

1. Materials Purchased by Broad Sectors of Industry
2. Output of Broad Sectors of Industry
3a Commodities Produced in the United Kingdom
3b Commodities Wholly or Partly Imported into the United Kingdom.

The first two relate to materials used or produced by large industrial groups; the last two are more detailed tables for particular commodities. Table 19.1 shows a typical example of the first table. It will be noticed that all groups are clearly defined with reference to the Standard Industrial Classification, also that the last two months' figures are provisional.

Prices are collected for over 10,000 materials and products, mostly from trade associations and large firms. All prices are carefully defined—grade, size or quality, terms of contract, etc. The index for any individual commodity is its current home-market price expressed as a percentage of its average home-market price in the base year, at present 1963. Where an index relates to more than one product (which it usually does), it is a weighted arithmetic mean of the price relatives, the weight being roughly proportional to the total values of the materials sold or purchased in 1963. Indices for the larger groups are calculated with "net" weights, i.e. they exclude sales from one firm to another in the same sector to avoid duplication.

The Wholesale Price Indices are widely used in business contracts, in forecasting future costs, and as factors for reducing current values to base year prices.

Index of Retail Prices

The history of this index goes back to the First World War, when the old Cost of Living Index Number was compiled in order "to measure the percentage increase month by month in the cost of maintaining unchanged the standard of living prevailing among working-class households in July 1914." No allowance was made for any changes in the standard of living since that date, and no attempt was made to measure the cost of living of the middle and upper classes.

Price quotations were collected from retailers, local authorities, etc., much as they are now, and each quotation was converted into a price relative based on July 1914. Indices for the five main groups

Table 19.1 Wholesale Price Index Numbers
(Extract from *Trade and Industry*)

1968 SIC	1969 3rd qr	1969 4th qr	1970 1st qr	1970 2nd qr	1970 3rd qr p	1969 Oct	1970 Mar	1970 Apr	1970 May	1970 June	1970 July	1970 Aug	1970 Sept p	1970 Oct p
MATERIALS PURCHASED BY MANUFACTURING INDUSTRY														
Orders III to XIX — Basic materials and fuel used in manufacturing industry:	121·4	124·1	126·3	126·1	124·9	122·8	127·0	126·8	126·4	125·3	125·0	123·9	125·7	125·5
Basic materials	121·8	124·7	126·7	126·5	124·9	123·2	127·4	127·2	126·8	125·5	125·1	123·7	125·8	125·8
†Fuel	119·8	124·9	124·0	124·0	124·8	120·9	124·7	124·0	124·0	124·0	124·4	124·7	125·1	125·1
Orders IV to XIX — Basic materials and fuel used in manufacturing industry, other than food, drink and tobacco manufacturing industries	126·1	127·3	130·1	129·2	129·9	125·9	130·8	130·4	129·3	128·0	127·5	126·1	127·1	126·0
Basic materials other than fuel	127·5	128·7	131·5	130·5	127·4	127·0	132·3	131·9	130·6	128·9	128·3	126·4	127·6	126·2
MATERIALS AND FUEL PURCHASED BY SELECTED BROAD SECTORS OF INDUSTRY														
211–229 Food manufacturing industries	112·4	118·0	119·4	120·8	121·6	116·8	120·4	120·4	121·2	120·6	120·9	120·5	123·3	124·5
Order V Chemicals and allied industries	116·8	118·7	120·7	122·4	123·7	117·5	121·9	122·1	122·3	122·8	123·5	123·4	124·2	125·5
311 and 312 Steel industries	116·9	118·8	123·3	125·2	125·7	117·8	124·4	124·1	125·7	125·8	126·2	126·4	130·4	130·1
Order VII (excluding MLH 342) Mechanical engineering industries	126·8	130·7	136·6	138·8	137·2	129·0	138·8	139·6	139·0	137·8	137·3	137·0	137·3	137·7
Order IX Electrical engineering industry	149·5	153·3	158·7	159·1	150·7	149·6	162·1	162·8	159·4	155·0	153·0	149·9	149·1	147·4
361 Electrical machinery industry	137·5	141·6	147·5	150·6	147·6	139·2	150·4	151·5	150·9	149·4	148·6	147·2	147·0	146·7
Order XIII Textile industries	103·4	102·8	103·6	104·2	104·3	102·8	103·6	103·8	104·3	104·4	105·0	104·4	103·6	102·6
Order XV Clothing and footwear industries	116·3	116·8	117·6	119·9	121·5	116·7	117·7	119·4	119·9	120·5	121·2	121·6	121·7	121·8
Order XVII Timber industries	127·8	129·3	131·6	134·0	135·6	128·8	132·6	133·4	134·1	134·5	135·1	135·6	136·0	136·2
481–484 Paper industries	121·8	123·6	129·4	131·5	135·5	122·7	130·0	130·6	131·6	132·1	135·0	135·6	135·8	136·0
Order XX Construction materials	119·9	121·9	124·9	128·2	130·3	120·7	126·2	127·4	128·2	128·9	130·0	130·5	130·5	131·7
Part of Order XX House building materials	122·3	124·1	127·1	130·2	132·9	122·9	128·3	129·3	130·1	131·2	132·4	133·1	133·2	133·9

†Coal (except for carbonizing), gas and electricity. Crude oil and coal for carbonizing are included in basic materials.

and for All Items were compiled from weighted arithmetic means
of the price relatives, the groups and their weights being as follows—

Group	Weight
Food	60
Rent	16
Clothing	12
Fuel and light	8
Other items	4
Total	100

This weighting was broadly based on the average expenditure shown
by 1944 working-class household budgets in a survey conducted by
the Board of Trade as far back as 1904. Only the bare necessities of
life were taken into account; it was assumed that the working classes
did not drink or smoke, or at least that they had no business to!

As standards of living improved, the Cost of Living Index became
seriously deficient. In 1937–1938, a new family budget inquiry was held,
but the war prevented the results being published and utilized. For
the next few years, the prices of food, rent and other basic necessities
were severely controlled, and since they were heavily weighted in
the Index, it showed a much smaller increase in the cost of living
than was really the case. It was completely discredited by June 1947,
when it was superseded by the present index.

The Index of Retail Prices is compiled by the Department of
Employment and published monthly in the *Department of Employment Gazette* and the *Monthly Digest of Statistics*. The change of
title is, as we might say in another connexion, highly significant.
Whereas the old Cost of Living Index had only covered the necessities
of life, the present index covers all normal items of expenditure on
retail goods and services. Luxuries and "conventional necessities"
have acquired greater importance, not only because of the greatly
improved standard of living of the working classes, but because the
coverage of the index has been extended to most small and medium
salary earners (see below). Nevertheless, the Index of Retail Prices
is still generally referred to in the Press and elsewhere as the "Cost
of Living Index," in spite of the Department's protestations that
they are not attempting to measure the cost of living.

The original series was called the Interim Index of Retail Prices,
because it was only intended as a stop-gap. It was six years, however, before another household budget survey took place, and it was
not until 1956 that that the new Index of Retail Prices appeared
without the "Interim."

The Household Expenditure Inquiry held in 1953, on which the second series was based, covered 13,000 households of all kinds, a representative sample of the United Kingdom. For the Index, however, two classes of household were excluded, viz. those of which the head had a gross income of £20 a week or more, and those in which at least three-quarters of the total income consisted of National Insurance retirement pensions and/or National Assistance. The remaining 11,638 family budgets provided a picture of the average pattern of expenditure of nearly all wage earners and most small and medium salary earners. As in the 1937–1938 inquiry, however, substantial adjustments had to be made for certain items, such as drink and tobacco, on which people spend more than they care to admit. Fortunately, reliable estimates for these items are available from Customs and Excise.

A continuous inquiry on a smaller scale, the Family Expenditure Survey, has been running since January 1957. Each year a new sample of about 10,000 addresses is selected, and the households living there are asked to keep detailed records of their expenditure for a fortnight and to supply information about certain regular payments over a longer period, and about incomes. About 70 per cent respond, yielding about 7,000 budgets. The results are published in an annual report. The weighting of the Index is now revised every year in accordance with consumption as ascertained from the Family Expenditure Survey for the three years ended the previous June, all items being revalued at the prices ruling at the date of revision.

Although the Family Expenditure Survey provides information on all kinds of payments, the Index naturally excludes such items as income tax, insurance premiums (other than house insurance) and National Insurance contributions, savings, subscriptions of all kinds and "investment" in football pools and other forms of betting. Capital payments and mortgage payments for house purchase are excluded as being investment rather than consumer expenditure, but the weight for housing includes an allowance for the "net rental equivalent" of owner-occupied dwellings.

At the time of writing the main groups and their weights, which are derived from expenditure during the period July 1966 to June 1969 adjusted to the prices ruling at 20 January 1970, are as follows—

Food	255
Alcoholic drink	66
Tobacco	64
Housing	119
Fuel and light	61
Durable household goods	60

Clothing and footwear	86
Transport and vehicles	126
Miscellaneous goods	65
Services	55
Meals bought and consumed outside the home	43

All items	1,000

Indices for these groups and for All items are weighted arithmetic means of price relatives based on 20 January 1970, and they are then multiplied by the corresponding indices for that date based on 16 January 1962. Example 19.1 illustrates the method with two groups only for simplicity.

Example 19.1

Find the Index of Retail Prices (All items) for 16 June 1970 (16 January 1962 = 100) given the index for 20 January 1970 and the indices for both months for Food and for All other items.

	16 Jan 1962 = 100		20 Jan 1970 = 100	Weights	Product
	20 Jan 1970	16 June 1970	16 June 1970		
Food	134·7	141·6	105·12	255	26,806
Other Items	135·8	139·4	102·65	745	76,474
All Items	135·5			1,000	103,280

Index for 16 June 1970 (20 Jan 1970 = 100) = 103·28

∴ Index for 16 June 1970 (16 Jan 1962 = 100)
$$= 103·28 \times 135·5$$
$$= 139·9$$

Prices are collected in various ways. Food prices are collected from retailers by local officers of the Department of Employment in 200 selected areas. Prices of beer and tobacco are obtained from the manufacturers, rent and rates from Local Authorities, transport from British Rail, London Transport, etc., and so on. Every effort is made to ensure comparability and allow for changes in quality, but items like furniture and second-hand cars present obvious difficulties.

Clearly it is necessary to take a relatively few prices as representing a whole group. For example, nine different kinds of furniture are listed and the price index for furniture is the average of the price relatives of these nine "indicators." The weight assigned to Furniture is, however, based on total expenditure by households on all kinds of furniture.

Special indices have recently been issued for 1-person and 2-person pensioner households, defined as householders where at least 75 per cent of the total income consists of National Insurance pensions, public assistance, etc. They exclude housing, and the weights are based on the expenditure of pensioner households. The indices are slightly higher than the general index.

Index Numbers of Agricultural Prices

Price indices are compiled by the Ministry of Agriculture, Fisheries and Food for agricultural products (crops, livestock, fruit and vegetables, etc.) and materials (fertilizers, feeding-stuffs, etc.). For subsidized materials such as fertilizers and lime, the prices used are market prices less subsidy, i.e. the final cost to the farmer. The present base is the period from July 1964 to June 1967 inclusive.

Consumers' Expenditure: Index Numbers of Prices

The Blue Book on National Income and Expenditure contains two tables of consumers' expenditure: (i) at current prices, (ii) at the prices ruling in a particular year, at present 1963. From these figures a system of price indices can be derived, more broadly based than the Index of Retail Prices, since it covers all consumers, and not merely those in a certain income group. The index for all consumer goods and services is, in fact, shown in another table. It differs from the Index of Retail Prices in one important respect, because it relates to current expenditure, i.e. it uses current weighting, while the Index of Retail Prices uses base-period weighting. This is less important now that the latter is revised every year.

Comparison with Pre-War Years

It is often required to compare the present value of the pound with its pre-war value, or to construct a continuous series of retail price indices over a long period. It may therefore be useful to construct such a series for the period 1914–1970. It is necessary to divide this period into three, linking up the component series by multiplication.

(i) For 1914–1938 the only available series is the old Cost of Living Index Number. Over this period it is probably fairly reliable. It is necessary to take July 1914 as the average for the year.

(ii) For 1938–1947 it is better to use the index numbers given or implied in the White Papers (as they then were) on National Income and Expenditure for 1946 and 1947.

(iii) For 1947 onwards the official Index of Retail Prices may be used, taking the index for 1947 as 100, although 17 June was the actual base date.

Strictly speaking, this is not one continuous series but the product of several.

The complete series, compiled in this way, is given in Table 19.2 with 1938 as base year.

Table 19.2   UK Index of Retail Prices, 1914–1970 (1938 = 100)

Year	Index	Year	Index	Year	Index
1914	64	1934	90	1954	240
1915	79	1935	92	1955	250
1916	94	1936	94	1956	263
1917	113	1937	99	1957	273
1918	130	1938	100	1958	281
1919	138	1939	102	1959	282
1920	160	1940	119	1960	285
1921	145	1941	130	1961	295
1922	117	1942	139	1962	308
1923	112	1943	143	1963	314
1924	112	1944	146	1964	324
1925	113	1945	149	1965	339
1926	110	1946	153	1966	353
1927	107	1947	168	1967	362
1928	106	1948	181	1968	379
1929	105	1949	186	1969	399
1930	101	1950	192	1970	425
1931	94	1951	201	1971	465
1932	92	1952	228		
1933	90	1953	235		

Suggestions for further reading

Monthly Digest of Statistics
Economic Trends
Trade and Industry
Department of Employment Gazette
Reports on the Family Expenditure Survey
Method of Construction and Calculation of the Index of Retail Prices
PHILLIPS, H. S., "United Kingdom Indices of Wholesale Prices, 1949–1956," *J. Roy. Stat. Soc.*, Series A, Vol 119, Part 3 (1956)

20

Labour Statistics

Published Sources

THE chief source of labour statistics in the United Kingdom is the *Department of Employment Gazette* (formerly the *Ministry of Labour Gazette*), published monthly. Until mid-1969 a quarterly booklet entitled *Statistics on Incomes, Prices, Employment and Production* provided a wide range of detailed tables, but this has been discontinued. As compensation there is a *Yearbook of Labour Statistics*, with a *Historical Abstract* giving series of figures up to 1968. Summary tables also appear in the *Annual Abstract of Statistics* and the *Monthly Digest of Statistics*. *Guides to Official Sources No. 1, Labour Statistics*, gives a very useful account of the statistical material both before and after the introduction of the National Insurance Scheme in 1948. For international comparisons there is the *Yearbook of Labour Statistics* and the monthly *International Labour Review*, both published by the International Labour Office.

The *Department of Employment Gazette* (hereafter abbreviated to "*D of E Gazette*") contains four sections. (1) Special Articles, often containing statistics, (2) News and Notes, (3) Monthly Statistics, consisting mainly of tables for the latest month, and (4) Statistical Series, giving a large number of time series on the following topics—

Working population
Employment
Unemployment
Unfilled vacancies
Hours worked
Earnings and wage rates
Retail prices

Industrial stoppages
Output per head and labour costs

Tables (other than regional tables) relate sometimes to Great Britain, sometimes to the United Kingdom.

Employment

At the time of writing, the Department of Employment derives its current statistics, particularly of employment, from two main sources. First, there is the annual count in June of the insurance cards used for National Insurance contributions. These cards last for twelve months and are exchanged for new ones, roughly one-quarter at a time, in March, June, September and December. From those exchanged in June a detailed analysis is made of the number of employees, by sex, industry and region, including those unemployed or temporarily absent from work. Separate figures are obtained for young people under 18, since their rates of contribution are lower than those of adults.

This analysis is supplemented by further information about age and other matters obtained from a 1 per cent sample. The additional items obtained for each person in this sample are date of birth, marital status (for women only), and, for both the current year and the previous year, the industry in which the person was employed and the region in which his or her card was exchanged. Information is thus provided about the age-distribution of numbers employed in each industry, numbers of married women employed, and movement between industries. The figures are, of course, subject to sampling errors which can themselves be estimated.

The second main source of labour statistics is the various "L" returns rendered at regular intervals by employers in manufacturing, distribution, and certain other industries and services, covering the great majority of employees. One such return shows the numbers on the payroll at the beginning and end of each month. By comparing the two sets of industry totals it is possible to obtain good estimates of relative changes during the month, and thus to project forward from month to month the total number employed in each industry, starting from the June figures based on the cards. After each annual analysis, the estimates for months following the previous June are revised.

The June analysis of cards includes employers and self-employed persons who are insured, but it is not possible to project these figures each month. The only firm totals of all employers and self-employed persons are provided by the Census of Population.

In June 1970, following the Government's decision to abolish National Insurance cards, a Census of Employment was taken to ensure that the necessary statistics could be obtained in this way when the cards ceased to exist. The form was, in fact, headed "Annual Census of Employment." At the time of writing, however, it is not known whether the change of government will affect the Department's plans.

The tables on employment published in the *D of E Gazette* fall broadly into two groups, annual and monthly, corresponding to the two main sources described above. The annual tables include an industrial analysis of employees (including unemployed) in Great Britain and in the United Kingdom at the previous June, and various tables based on the 1 per cent sample, such as an analysis of employees in Great Britain by age and region, or by age and industry. There is also an annual article entitled "Young Persons Entering Employment," giving various analyses by sex, age at entry, type of employment, industry group and region.

The monthly tables include analyses by industry and by region of employees in employment, as well as a summary of the *working population*. This is not, as one might think, the number at work, but rather what an actuary would call the number exposed to risk of work. The working population of Great Britain in March 1971, for example, was made up as follows—

Employees in employment	21,970
Employers and self-employed	1,744
HM Forces and Women's Services	369
Wholly unemployed	700
Total working population	24,783

(*Source: D of E Gazette*, Jan 1972, Table 101)

The numbers employed include part-time workers, i.e. persons who normally work for not more than 30 hours a week, but a separate table is published once a quarter for women employed part-time in manufacturing industry, analysed by industry.

The warnings given earlier about breaks in continuity and the need for careful definition apply particularly to employment statistics. The worst break occurred in 1948, when the National Insurance Scheme was introduced. In the old series, which was based on numbers insured against unemployment under the Unemployment Insurance Acts, the total in civil employment at mid-1948 was 19,064,000. In the new and more comprehensive series it was 21,569,000. The increase was due mainly to—

(*a*) the inclusion of about half a million indoor domestic servants and nearly a million workpeople over pensionable age (65 for men, 60 for women), both of which classes were previously excluded; and

(*b*) the counting of about 800,000 part-time workers as full units instead of half units.

Another break followed the revision of the Standard Industrial Classification in 1958, although the Ministry provided a good link by showing the composition of the working population for several months of 1959 on both old and new classifications. The most important changes were—

(i) The transfer of certain trades (particularly motor repairing and garages, and boot and shoe repairing) from Manufacturing Industries to Services.

(ii) The regrouping of industries, particularly in the Engineering and allied industries.

A further revision of the SIC in 1968 caused only minor changes.

Unemployment

Unemployment statistics are derived from monthly returns rendered by Employment Exchanges and Youth Employment Service Careers Offices, showing particulars of the unemployed persons on their registers on a Monday in each month. They include not only persons who have fallen out of employment, but school-leavers and others registering for their first employment, so there is usually a seasonal increase in the number registered as unemployed at the end of the school summer term. Persons working on short time or temporarily stood off by their employers are included if registered (registration at an Employment Exchange is a condition for receiving Unemployment Benefit under the National Insurance scheme), but persons already in employment who wish to change their jobs are of course excluded.

The details obtained each month from these returns are an industrial analysis by sex, adult or juvenile (under 18), and whether wholly unemployed or temporarily stopped. There is a broad analysis of those wholly unemployed by duration (see Table 20.1), and other minor details. A more detailed analysis by age and by duration is taken in July and January, and by duration in April and October also. Also, each quarter, an analysis by occupation as well as industry is taken of adults wholly unemployed and of the numbers of vacancies unfilled.

Since unemployment is a vital economic indicator—and sometimes political dynamite—it is important that the significance of

Table 20.1 Registered wholly unemployed in Great Britain(1)

(Monthly Digest of Statistics, June 1970)

Analysis by duration of unemployment

Thousands

	Males				Females			
	Total (including school-leavers) (2)	2 weeks or less	Over 2 but not more than 8 weeks	More than 8 weeks	Total (including school-leavers) (2)	2 weeks or less	Over 2 but not more than 8 weeks	More than 8 weeks
1967 Monthly	416·7	72·5	102·4	241·9	100·1	22·5	29·1	48·5
1968 averages	457·3	73·6	107·7	276·0	88·6	19·7	25·5	43·5
1969	459·4	76·4	109·9	273·1	81·7	19·4	24·3	38·0
1969 September 8	452·7	75·5	111·1	266·1	85·0	21·4	26·8	36·9
October 13	453·6	85·0	113·5	255·2	86·4	24·0	28·0	34·4
November 10	463·7	80·5	118·5	264·8	85·8	20·5	28·8	36·5
December 8	480·5	76·9	120·7	282·8	82·3	16·3	25·7	40·3
1970 January 12	523·6	89·9	131·2	302·6	85·1	20·7	23·5	41·0
February 9	517·4	80·8	121·2	315·4	86·1	19·2	24·9	41·9
March 9	514·3	77·7	121·1	315·5	84·6	17·6	25·4	41·6
April 13	505·6	85·2	114·0	306·3	85·0	20·7	24·0	40·4
May 11	470·8	70·8	103·8	296·2	79·8	16·1	22·4	41·3
June 8	448·0	70·0	93·6	284·3	73·2	15·6	19·0	38·7
July 13	467·7	87·8	111·2	268·7	81·2	22·4	22·5	36·2
August 10	499·5	81·7	131·7	286·1	95·5	22·4	33·0	40·1
September 14	485·0	87·1	108·8	289·1	92·1	24·6	26·6	41·0
October 12	480·9	84·8	119·1	277·0	93·0	24·7	30·2	38·0

Source: Department of Employment.

(1) Excluding casuals.   (2) School-leavers are persons under the age of 18 registered as unemployed who have had no insured employment since completing full time education.

changes in numbers unemployed should be correctly assessed. At certain times of the year, e.g. just after Christmas and at the end of school terms, there are strong seasonal effects. Many of the figures are therefore given both actual and seasonally adjusted.

Labour Turnover

Each quarter the *D of E Gazette* publishes a table of labour turnover rates in the principal manufacturing industries. The only important omission is shipbuilding.

The basic data are obtained from the "L" returns already mentioned. They relate to periods of four weeks, and include administrative and clerical employees as well as operatives. As "L" returns are only required from firms employing over ten persons, smaller firms are not represented.

Persons who are engaged during the period but leave before the end of it are not included; this is not unimportant, for a large number leave within the first few weeks. Losses are calculated from the numbers employed at the beginning and end of the period and the number of engagements, and therefore tend to be understated to the same extent as engagements. On the other hand, losses include persons who have merely changed their employers within the same industry, and the percentage of losses therefore exceeds the total wastage from the industry as a whole.

Overtime and Short-time

Two other useful guides to the state of the nation's economy are the amounts of overtime and short-time being worked. Figures are collected from all establishments with more than ten employees in manufacturing industries other than shipbuilding and ship-repairing. They relate only to operatives and exclude administrative, technical and clerical staff.

Distinction between Wage Rates and Earnings

Most manual workers are paid either time rates, a given amount for a unit of time such as an hour or a standard working week with bonuses or higher rates for overtime, or piece rates, a given amount for each unit of work performed. An annual volume entitled *Time Rates of Wages and Hours of Work* sets out for the more important industries the minimum or standard time rates of wages as fixed either by voluntary agreements between organizations of employers and workpeople or by statutory orders. In general, however, wage rates are less important than earnings. "Earnings" are the worker's pay-packet, his actual pay for the week or other period, including bonuses and overtime, but before deducting income-tax, insurance contributions, etc.

In comparing wage rates or earnings in different industries, or for different groups of workers, it must be remembered that cash earnings are often supplemented by some form of payment in kind or "concealed emoluments," not to mention anything so obvious as tips. Clerical staff may receive luncheon vouchers, railway employees are allowed free or cheap travel, and so on. Some pension schemes are contributory and others are non-contributory. Comparisons should also take into account the proportions of men and women, skilled and unskilled workers, etc., in the groups or industries concerned.

Index Numbers of Wage Rates and Hours Worked

Indices of basic weekly and hourly wage rates and hours of work appear monthly in the *D of E Gazette* and the *Monthly Digest of Statistics*. One item, the "Index of Weekly Wage Rates," has a long history for which the interested reader should refer to an article on "Official Indices of Rates of Wages, 1880–1957" in the *Ministry of Labour Gazette* for April 1958.

All three series, at the time of writing, are based on 31 January 1956, a base-date purposely chosen close to that of the Index of Retail Prices (which was then 17 January 1956). In the Index of Basic Weekly Rates of Wages, the figures for All Industries and for industry groups are weighted arithmetic means of the indices for their component trades, the weights being approximately proportional to the total wage bills of those trades in October 1955. These were estimated by multiplying the average earnings for each group as shown by the Inquiry into Earnings and Hours of Wage-Earners (see below) by the number of manual workers in that group. The weighting system is thus constant throughout the present series.

Indices of "normal" weekly hours are derived from the information received about changes in the basic working week. These indices are then divided into the indices of weekly wage rates to obtain indices of hourly rates. Thus, taking the figures for 1969 shown in Table 20.2,

$$177 \cdot 6 \div \frac{90 \cdot 6}{100} = 196 \cdot 0$$

Table 20.2 shows the three series for men (All Industries and Services) for a run of years. The last column shows the effect of the movement for a shorter basic working week, which began in mid-1959 and has generally resulted in the basic week being reduced by four hours. The average number of hours worked has not been reduced by this amount. It simply means that more time has been counted as overtime.

Table 20.2 Indices of rates of wages and normal weekly hours
for men—all industries and services
(31 January 1956 = 100)

Year	Basic weekly rates of wages	Basic hourly rates of wages	Normal weekly hours
1956	104·8	104·8	100·0
1957	110·0	110·1	99·9
1958	113·8	114·2	99·7
1959	116·8	117·3	99·6
1960	119·7	122·3	97·9
1961	124·6	129·8	96·0
1962	129·1	135·7	95·1
1963	133·6	140·6	95·0
1964	139·8	147·8	94·6
1965	145·7	156·9	92·9
1966	152·2	167·0	91·1
1967	157·9	173·8	90·9
1968	168·6	185·9	90·7
1969	177·6	196·0	90·6
1970	195·2	215·9	90·4

Earnings and Hours of Wage-earners

Inquiries into weekly earnings, as distinct from wage-rates, of manual wage-earners were made in 1886, 1906, 1924, 1928, 1931, 1935 and 1938, sometimes but not always accompanied by inquiries into hours of work.

The present series of inquiries, now carried out every October, dates from July 1940, when the Ministry of Labour (now the Department of Employment) asked a large number of employers for particulars of numbers employed and wages paid. Since July 1943 particulars of hours worked have also been required. A form is shown on page 236 which shows the information required by the Department. The student should study it carefully, noting particularly how wage-earners at work, wages paid and hours worked are defined in the Notes for Guidance (reproduced on page 237).

At the time of writing, the inquiry, which is still voluntary (unlike other inquiries by the D of E), covers nearly six million workers in about 50,000 establishments. The principal employments excluded are agriculture, coal-mining, railways, shipping, distribution and catering trades, commerce and banking, entertainment and domestic service. Statistics of earnings in agriculture, railways, and coal-mining, however, are compiled by the relevant bodies, such as the National Coal Board.

Name of Firm...

Address of Works covered by this Return

....

....

(If possible a separate Return should be made for each Works or Establishment)

Telephone number ....

Most important product (or activity)
carried on at above address ....

Full-time workers

(Figures relating to part-time workers, i.e. workers whose employment ordinarily involves service for not more than 30 hours a week, should be shown in columns (4) to (6))

M/C line No.		Numbers of manual workers, EXCLUDING part-time workers, at work in the pay-week which includes 7 October 1970 See Notes (1), (2) and (3)	Total wages (rounded to nearest £) paid to these workers for the specified week See Notes (4), (5) and (6)	Total number of hours (rounded to nearest hour) actually worked by these workers for the specified week See Note (7)
		(1)	(2)	(3)
			£	
1	Men (21 years and over)			
2	Women (18 years and over)			
3	Youths and Boys (under 21 years)			
3	Girls (under 18 years)			
4	Total			

Part-time workers

If, in the pay-week, you were employing any men or women as part-time workers (as defined in previous columns), please state—

M/C line No.		Numbers of part-time workers	Total wages (rounded to nearest £) paid to these part-time workers for the specified week	Total number of hours (rounded to nearest hour) actually worked by these part-time workers for the specified week
		(4)	(5)	(6)
			£	
1	Men (21 years and over)			
2	Women (18 years and over)			
4	Total			

Notes for Guidance

1. **The pay-week.** If your works were stopped for a part or the whole of the pay-week specified because of a general or local holiday, breakdown, fire or industrial dispute, please give figures for the nearest week of an ordinary character.

2. **Classes covered.** The particulars given should relate to manual workers who were at work during the whole or part of the week, *excluding* workpeople doing work at home on material supplied by the employer. Foremen (except works foremen), transport workers, warehousemen, canteen workers (if employed by you), etc. should be *included*, but administrative, technical and office employees, commercial travellers, shop assistants and canteen workers employed in canteens conducted by the employees themselves or by independent contractors should be *excluded*. (Works foremen are covered by the annual S.L. (Administrative, Technical and Clerical staff) inquiry form.) If any workpeople employ helpers, those helpers should be *included*.

3. **Workpeople absent.** Workpeople absent during the whole of the pay-week (except those who are available for work under a guaranteed wage agreement) should be *excluded*; those at work during any part of the pay-week should be *included*.

4. **Wages paid.** These figures should show the total gross earnings of the workpeople included in this return for the specified pay-week, inclusive of overtime, etc., *before any deductions are made (for income tax, workers' insurance contributions, etc.)*. Payments made under a guaranteed wage arrangement should be *included*.

5. **Bonuses** (including non-contractual gifts and bonuses). These should be *included*. Where a bonus is paid otherwise than weekly (e.g. half-yearly), the proportionate weekly amount of the bonus should be included; if the amount of the current bonus is not known, the amount paid for the previous bonus period should be used for this calculation.

6. **Employers' contributions to National Insurance, holiday funds, pension schemes, etc.** These should *not* be counted as wages.

7. **Hours worked.** The figures to be entered are the aggregate hours worked, *including* overtime, e.g. for men, the total of all the hours worked by the men included in columns (1) or (4). Time lost, e.g. through short-time working or absence from work, should be *excluded*, except that any hours during which workpeople were available for work and for which a guaranteed wage was paid to them should be counted as hours actually worked. *If any workpeople have been included in column* (1) *or column* (4) *whose actual working hours are not known, their estimated working hours should be* included. Hours paid for but not worked by young persons on day release schemes should be *included*.

Where overtime has been paid for at rates above the ordinary hourly rates (e.g. time and a quarter, time and a half, etc.), the figures given should be the hours actually worked and not the number of "pay-hours" (e.g. if 4 hours overtime were worked and paid for at time and a quarter, the number of hours to be included in the figure should be 4, and not 5).

The results of the inquiry are published in two tables covering four closely-printed pages and showing, for each industry and for each category of workers, i.e. men (21 and over), youths and boys, women (18 and over), and girls—

(a) Number of wage-earners covered by the returns.
(b) Average earnings in the week in question.
(c) Average number of hours worked in that week.
(d) Average hourly earnings in that week.

Separate figures are shown for full-time and part-time women employees.

The Department of Employment is careful to point out that: "In view of the wide variations, between different industries, in the proportions of skilled and unskilled workers, in the opportunities for extra earnings from overtime, night-work and payment-by-results schemes and in the amount of time lost by short-time working, absenteeism, sickness, etc., the differences in average earnings shown in the tables should not be taken as evidence of, or as a measure of, disparities in the ordinary rates of pay prevailing in different industries for comparable classes of workpeople employed under similar conditions."

The October inquiry yields only broad average figures of earnings and hours. In order to obtain information about the distribution and composition of earnings of different groups of employees, and related matters such as overtime, sick pay and absenteeism, a 1 per cent sample survey of employees was carried out in April 1970. This also will probably become an annual inquiry.

Wage Drift

There is a continual tendency for average earnings to increase relatively faster than basic wage rates, owing to demand for labour and consequent increases in overtime and bonuses. This is known as *wage drift* or, in some quarters, wage creep. Table 20.3 shows it as the excess of the percentage increase in one year in average hourly earnings (excluding overtime) over the percentage increase in the Index of Hourly Wage Rates.

Other Statistics of Earnings

A somewhat rough and ready Index of Average Earnings is compiled monthly, covering both weekly and monthly paid employees. Monthly salaries are converted to weekly equivalents, then the total gross remuneration is divided by the total number of employees. There are considerable fluctuations and the index has met with some criticism.

Twice a year tables are published for Occupational Earnings in Engineering and other Industries, giving average earnings, both including and excluding overtime, for various trades and categories of workers. Shift allowances are included, so that changes in the proportions of day and shift workers may vitiate comparisons.

Information about salaries is scanty compared with that on wages. There is, however, an annual "SL" Inquiry into earnings of administrative, technical and clerical employees, covering all salaried staff in a wide range of industries and services. The results are published in the form of tables of average earnings of monthly-paid and weekly-paid males and females, for each October covered by the inquiries, both in money and in index form. It is understood that this annual inquiry will shortly be discontinued.

Clerical Salaries Analysis

The survey with the above title is carried out every even year, not by the Department of Employment but by the Institute of Office Management. The fourteenth survey, taken in 1970, covered about 70,000 clerks in 700 offices. Employers are asked to analyse their clerks by six carefully defined grades and 35 weekly salary ranges. The report consists principally of Area Tables giving, for each region, large town, or other well-defined area, the median and

Table 20.3 Wage drift: percentage changes over corresponding month in previous year: United Kingdom

(*Employment and Productivity Gazette*, July 1970)

	Average weekly wage earnings (1)	Average hourly wage earnings (2)	Average hourly wage earnings excluding the effect of overtime* (3)	Average hourly wage rates (4)	"Wage drift" (col. (3) minus col. (4)) (5)
1956 April	+8·6	+9·1	+9·3	+8·3	+1·0
October	+7·3	+7·9	+8·2	+7·6	+0·6
1957 April	+3·5	+3·6	+3·8	+2·5	+1·3
October	+5·8	+6·5	+6·6	+5·6	+1·0
1958 April	+4·6	+5·5	+5·9	+4·8	+1·1
October	+2·3	+3·1	+3·4	+3·7	−0·3
1959 April	+3·9	+3·6	+3·5	+3·5	−0·0
October	+5·1	+3·6	+2·9	+1·4	+1·5
1960 April	+6·5	+7·0	+6·4	+4·4	+2·0
October	+6·6	+8·1	+7·3	+5·5	+1·8
1961 April	+6·6	+7·3	+6·5	+6·2	+0·3
October	+5·4	+7·0	+6·9	+6·4	+0·5
1962 April	+4·0	+5·1	+5·2	+4·1	+1·1
October	+3·2	+4·1	+4·4	+4·2	+0·2
1963 April	+3·0	+3·6	+4·0	+3·6	+0·4
October	+5·3	+4·1	+3·6	+2·3	+1·3
1964 April	+9·1	+7·4	+6·5	+4·9	+1·6
October	+8·3	+8·2	+8·1	+5·7	+2·4
1965 April	+7·5	+8·4	+8·0	+5·3	+2·7
October	+8·5	+10·1	+9·5	+7·3	+2·2
1966 April	+7·4	+9·8	+9·7	+8·0	+1·7
October	+4·2	+6·2	+6·5	+5·6	+0·9
1967 April	+2·1	+2·8	+3·0	+2·7	+0·3
October	+5·6	+5·3	+5·0	+5·3	−0·3
1968 April	+8·5	+8·1	+7·7	+8·6	−0·9†
October	+7·8	+7·2	+7·0	+6·7	+0·3
1969 April	+7·5	+7·1	+6·9	+5·4	+1·5
October	+8·1	+8·0	+8·0	+5·5	+2·5

Note—
The table covers all full-time workers in the industries included in the department's half-yearly earnings inquiries (Table 122).
* The figures in column (3) are calculated by—
1. Assuming that the amount of overtime is equal to the difference between the actual hours worked and the average of normal weekly hours;
2. Multiplying this difference by 1½ (the assumed rate of overtime pay);
3. Adding the resultant figure to the average of normal weekly hours to produce a "standard hours equivalent" of actual hours worked; and
4. Dividing the average weekly earnings by the "standard hours equivalent" which gives a reasonably satisfactory estimate of average hourly earnings exclusive of overtime.
† The negative wage drift was mainly due to the special factors arising from implementation of the later stages of the December 1964 long-term national agreement for the engineering industry.

quartile salaries of each grade of clerk. Care is taken to avoid spurious accuracy, the figures being rounded and shown only for areas with more than six establishments employing a total of 50 clerks or more.

References and suggestions for further reading

Department of Employment Gazette
Yearbook of Labour Statistics
British Labour Statistics, Historical Abstract **1886–1968**
Monthly Digest of Statistics
Annual Abstract of Statistics
Guides to Official Sources No 1: Labour Statistics
Year Book of Labour Statistics (ILO)
Monthly Labour Review (ILO)
Clerical Salaries Analysis (The Institute of Office Management)

21

Production

The Nature of Production Statistics

In this chapter we take "production" to mean the growth, extraction or manufacture of tangible goods, which can be counted or measured, as distinct from services, such as finance, education, transport and retail trade. Production can be measured at various stages; we can record the amount of coal mined at a colliery, the quantities of dyestuffs made from that coal, the quantities of yarn dyed with those dyestuffs and the number of tablecloths, curtains, etc. made with that yarn. In other words, we can measure output of raw materials, intermediate goods, and finished goods.

Often the finished goods of one industry are the raw materials of another. This can, if we are not careful, lead to double counting, but it is often convenient to measure output at different stages of a process, even within the same factory. Thus, the monthly return for "Chemicals (General)" issued by the Department of Trade and Industry asks for total production of sulphuric acid—not merely production for sale—so the figure supplied by a manufacturing firm will include acid used in further manufacture, e.g. of sulphate of ammonia.

A distinction is sometimes made between producers' goods and consumers' goods, but this is not watertight. Agricultural products are consumers' goods when eaten by human beings, but producers' goods when eaten by cattle. Cars may be producers' goods in working hours and consumers' goods during hours of leisure.

For many purposes it is convenient to deal with Industrial Production as a group, i.e. manufacturing industry, mining and quarrying, construction and public utilities, and to omit agriculture or treat it separately. For one thing, it is impossible to measure output of crops and livestock month by month or even quarterly in any meaningful way: the results of a single month's work cannot be estimated and may not even be visible until long afterwards. We shall therefore deal briefly with agriculture, then consider industrial production in more detail.

Agricultural Production

The first Census of Agriculture was taken in 1865, and a full census of holdings of more than one acre is taken every June. The main publication is the annual *Agricultural Statistics: United Kingdom*, which gives the results of the censuses, numbers of livestock, acreage and production of crops, etc. Figures of production and stocks of a wide range of agricultural products are given in the *Annual Abstract* and the *Monthly Digest of Statistics*. A large number of other publications containing statistics of agriculture, fisheries, etc. are issued in this and other countries, and by such bodies as the Food and Agricultural Organization of the United Nations.

Industrial Production: Statistical Sources

Apart from the annual Census of Production and the monthly Index of Industrial Production, both of which are discussed later in this chapter, there is a considerable volume of production statistics for individual products in the *Annual Abstract*, the *Monthly Digest of Statistics* and other publications, often accompanied by figures of consumption and stocks. *Trade and Industry*, for instance, gives production of man-made fibres, *Housing Statistics* gives quarterly figures of houses started, under construction and completed, and car production is detailed in *The Motor Industry of Great Britain*, published by the Society of Motor Manufacturers and Traders.

In 1962 the Board of Trade started a series of leaflets entitled *Business Monitor: Production Series*, published for the benefit of industry and available on subscription. Each series gives either monthly or quarterly statistics of production, sales, etc., in Great Britain for a particular commodity or group of products.

All this information is supplied by industry in the first place, and many firms make regular returns of production, stocks, etc., on a voluntary basis to the Department of Trade and Industry or other Government Departments, in addition to the returns made compulsorily in the Census of Production, now to be described.

The Census of Production

The Census of Production is at present the basis of the production statistics compiled and published by Government Departments. By providing figures of value of the output of different products, it provides the weighting system for the Index of Industrial Production (*see below*) and serves as a guide to the short-term statistics that should be collected. It also provides information of value for economic analysis, e.g. in studies of different industries and of the relations between industries.

In general, the Census covers undertakings in all fields of industrial production, including construction, gas, electricity and water supply, mines and quarries. The line between production and services, however, is somewhat arbitrary; for example, laundry, cleaning and dyeing trades were at one time included, but are now excluded. Repair work is generally covered by the Census of Distribution, but aircraft, ship and locomotive repairs are regarded as an essential part of industry and are included in the Census of Production.

The first Census was held in 1907, followed by others in 1912, 1924, 1930 and 1935. From 1948 onwards, there has been an annual census, full censuses being held in 1948, 1951, 1954, 1958, 1963 and 1968, and simpler sample censuses in other years. Among the full censuses, some have been more detailed than others: in 1951 and 1958, for example, firms were not asked for details of purchases of materials, stores and fuel, as they were on the other occasions.

Small firms, defined as those with an average of fewer than 25 employees during the year, need only answer summary questions about numbers employed, etc. In the 1968 Census, larger firms had to answer questions about—

D   Working proprietors
E   Employment
F   Wages and salaries, etc.
G   Stocks
H   Capital expenditure
J   Other selected items of expenditure
K   Purchases
L   Sales and work done
N   Total make and electricity generated.
Q   Classification of business.

Items K and L are spelled out in great detail. The Census might really be called a Census of Sales rather than of Production. Sales of the firm's products are required by net selling value and wherever possible by quantity. Net selling value is defined as the amount charged to customers, whether valued "ex works" or "delivered," less purchase-tax, trade discounts, agents' commissions, allowances for returnable cases, etc. Where goods are charged as delivered to customers overseas, the f.o.b. (free on board) value is required. Goods transferred from one department to another must be valued as far as possible as if they had been sold to an independent purchaser, and if the receiving department makes a separate return, such goods must be valued on the same basis in its Materials section, with due allowance for any payments to transport firms. Where two or more distinct trades are carried on in separate departments of a single works, separate returns are generally required.

The *gross output* of a trade is defined as the total value of goods made and other work done during the year. The *net output*, which in many ways is more important, is equal to the gross output less the cost of all purchased materials, fuel, etc., used, any payments for work given out to other firms or persons, any transport payments to other firms, and any Customs and Excise duty. In short, net output is the value added by manufacture to the raw materials. In the United States Census of Manufactures it is called "value added by manufacture," and in some countries simply "added value."

Table 21.1 shows how gross and net output were calculated for the Shipbuilding Industry in 1963. The figures, incidentally, are for "larger firms" only. We begin with sales of goods produced, including the value of work done for other firms, add the value of merchanted goods and canteen takings, and two other items which may be negative,

(i) the change in the value of work in progress, i.e. stocks of intermediate goods,
(ii) the change in the value of goods on hand for sale, i.e. stocks of finished goods.

All these constitute Gross Output. We then find the cost of materials used, starting with the cost of materials purchased and *subtracting* the increase in stocks of materials (if stocks have increased, we consumed less than we bought during the year). We deduct the result from Gross Output, together with payments to other firms for work they have done, and arrive at Net Output.

Both gross and net output, as defined here, exclude any purchase-tax, etc., levied on the finished products, and are valued before any deductions for subsidies. In other words, they are valued "at factor cost," i.e. at the prices paid to the various factors of production—labour, management, land, capital, etc., including profits, whether distributed or not. The Census Reports drive the point home by saying ". . . it (net output) constitutes the fund from which wages, salaries, insurance, pensions, hire of plant and machinery, payments for repairs and maintenance, costs of operating road vehicles, rents, rates and taxes, advertising and other selling expenses and all other similar charges have to be met, as well as depreciation and profits."

Table 21.1 Calculation of gross and net output in shipbuilding and marine engineering, 1963

	£'000
Sales of goods produced and work done	400,613
Merchanted goods and canteen takings	2,823
Change in work in progress	2,400
Change in goods on hand for sale	−288
Gross Output	405,548
Purchases of materials, fuel, etc.	155,846
Purchases of goods for merchanting and canteen purchases	2,544
add Decrease in stocks of materials, etc.	1,074
Cost of Materials, etc. used	159,464
Gross output	405,548
less cost of materials, etc. used	−159,464
less payments to other firms for work done and transport	−40,227
Net Output	205,857

The UK Index of Industrial Production

The index compiled for the United Kingdom by the Central Statistical Office is published monthly in *Trade and Industry* and in the *Monthly Digest of Statistics*. The present base year is 1963, so that the index for any month gives a comparison with the average month of 1963. Agriculture, trade, transport, finance and all other public or private services are excluded for reasons discussed above. In fact, it covers the same industries as the Census of Production.

Separate indices are published for each of the main industry groups in the Standard Industrial Classification, their weights being roughly proportional to the net output of the corresponding industries in 1963, as revealed by the Census of Production for that year. The index thus gives a comparison of the net output of the current month with that of the average month of 1963, both valued at 1963 "prices," price in this case meaning net output per unit of quantity.

The index is built up from production relatives of 892 individual series, the data being generally supplied by the industry concerned. Whenever possible they are based on output in tons or some other physical measure, but often this is impracticable. Sometimes it is necessary to use sales or deliveries instead of production. Failing that, "deflated value" may be used, i.e. value of sales or output deflated by a suitable price index, as in the case of clothing. Another method, which involves a little time-lag, is to use the quantity of coal, iron ore, or other raw material used in production, as in the case of newspapers and periodicals; this is unsatisfactory if the

9

proportion of the material used to make the product changes, as happened when filter-tipped cigarettes became more popular. Finally, as a last resort, it may be necessary simply to use numbers employed, as in water supply.

Heterogeneity is a constant problem. As with price indices, an index of production inevitably fails to take into account many changes in quality and variations in size, grade, etc. that might well vitiate the index if they were biased in one direction, e.g. if there were a tendency to produce larger motor-cars, better houses or stronger beer. In some cases, such as drugs, where there is infinite variety and new products are constantly appearing, it is impossible to give an estimate of volume of production with any confidence, just as it is impossible to compile a reliable price index for them.

Adjustments are made for variations in the number of working days in the month, so that in effect the indices compare the average weekly rates of production in different months, but they still reflect fluctuations due to public holidays, annual holidays and other seasonal influences. Monthly indices are therefore shown seasonally adjusted as well as unadjusted. The seasonal factors are recalculated each year.

Input-Output Tables

The relations between one industry and another may be illustrated in the form of tables showing sales by one industry to another and purchases by one industry from another. This is really a two-way analysis of the gross output of each industry group, etc., showing the various components of input and output of each group and arriving at the same total in both cases. For completeness these input-output tables, as they are called, must include sales to final consumers, exports, investment and stock appreciation on the one hand, and labour costs, profits, etc., on the other. Table 21.2 shows a simplified input-output table for a hypothetical country with only three industries.

The output of each industry group as shown in such tables is "free of duplication," i.e. it excludes all transactions between establishments within that group. It therefore differs from gross output as defined in the Census of Production, which includes sales by establishments to other establishments in the same industry.

Various other tables can be compiled from the original table, such as an analysis of the final output of each industry group into the percentages contributed by each group (including itself) in terms of net output, or a table showing what proportion of each industry's output is exported, directly or indirectly.

An analysis of this kind is only possible when there is a detailed

Table 21.2 A simplified input-output table

Sales by industry group	Purchases by industry group					
	Industry A	Industry B	Industry C	Final buyers	Exports	Total output (£ million)
Industry A	—	25	50	170	45	290
Industry B	40	—	90	35	15	180
Industry C	100	20	—	60	30	210
Imports	20	15	30	35	—	100
Taxes on expenditure *less* subsidies	5	—	10	15	—	30
Wages, salaries, etc.	90	80	25	—	—	195
Gross profits	35	40	5	—	—	80
Total input	290	180	210	315	90	1,085

Census of Production, providing figures of both sales and purchases by industry. That is why Input-Output Tables were published for 1954 and 1963, but not for 1958.

Output Per Head

A table with this heading appears in the *Monthly Digest of Statistics* and gives indices, for a few broad groups, of—

Output (identical with the Index of Industrial Production)
Employment
Output per head

The last item is simply the first divided by the second.

It is not proposed to discuss at great length what is primarily not a statistical, but an economic problem. It is necessary, however, to sound a warning about the uncritical use of ratios or indices purporting to measure productivity. The commonest of these is ratio of output, whether in physical or financial terms, to numbers employed or man-hours worked. This is often called "labour productivity" to distinguish it from other ratios involving capital employed, materials used, etc. It is a misleading term, as it suggests that changes in output per man are due to labour alone whereas they may well be due to increased mechanization, improvements in technique and other causes.

This is not to say that it is useless to study variations in output per man or per man-hour. In certain industries, e.g. cotton, some very useful work has been done on these lines, but the farther such calculations get from the factory floor, the more academic they become. Broad indices of output per worker for a whole firm or industry are of little practical value, and may be misleading in such matters as wage negotiations.

International comparisons must also be treated with caution. An industry or product may represent different things in different countries (think of the American "drug-store"). It would be unreasonable to compare the number of cars produced in the United States with the number of much smaller cars produced in the United Kingdom, and where the output of an industry can be expressed only in terms of money, rates of exchange may present difficulties. It may, however, be easier to compare the rates of change of productivity in different countries.

References and suggestions for further reading

Annual Abstract of Statistics
Monthly Digest of Statistics
Trade and Industry
Reports on the Census of Production
Input-Output Tables for the United Kingdom, 1963
Studies in Official Statistics, No 17, "The Index of Industrial Production and Other Output Measures"
Guides to Official Sources, No 6: "Census of Production Reports"
CARTER, C. F., REDDAWAY, W. B., and STONE, J. R. N., *The Measurement of Production Movements* (Cambridge University Press)

22

Internal Trade

We turn now from production to what might be called the rest of the process, i.e. the distribution and sale of goods to the consumer, including such services as catering, hairdressing, laundries and most repair work. We shall also include in this chapter brief sections on stocks, capital expenditure and orders. Until the mid-1950's statistics on all these items were scanty in the extreme, but they are now greatly improved. Regular figures for retail and catering sales, hire purchase, stocks, etc. are published in the *Monthly Digest of Statistics* and in *Trade and Industry*, but, as with production, the basis of all these statistics is the Census, now to be described.

The Census of Distribution

The Census of Distribution and Other Services, to give it its full title, provides valuable information for business firms, local authorities, market research workers, etc. and forms a framework for the Index Numbers of Retail Sales and other short-term statistics, and for any sample surveys that may be required. For details of the earlier censuses, the reader should refer to other books or the previous edition of this one, or to the Census Reports. Only a brief summary will be given here.

The first full Census of Distribution was taken for 1950, after a pilot sample census for 1947. The second full census was for 1961 and the third for 1971, sample censuses being taken for 1957 and 1966. This pattern seems likely to continue. Each census covers the whole of Great Britain, but detailed analysis by region and area is only possible with a full census.

The content of the census varies from time to time. The 1950 Census included wholesale trade, catering and the motor trade, each of which is now the subject of a separate quinquennial inquiry. The 1957 Census had questions about all forms of consumer credit and the largest retailers were asked about capital expenditure. The 1961 Census included laundries, launderettes and dry cleaners for the first time. For 1966, radio and television relay services were included.

Retail organizations are classified into three groups (see Table 22.1)—

Co-operative societies
Large multiple organizations (with ten or more branches)
Independent organizations

Certain other terms which are used in the Census Reports need to be defined—

(a) *Establishment:* a single shop or a single branch of a multiple or other organization.

(b) *Persons engaged:* not merely employees, but all who work in the business, including working proprietors, members of the owner's family and part-time workers.

(c) *Turnover:* total trading receipts before deducting any outgoings.

(d) *Gross margin:* turnover less purchases plus increase (or less decrease) in stocks during the year. Rather like net output, but making no deduction for fuel and other goods, such as display materials, which are used, not sold.

(e) *Stockturn:* the ratio of annual turnover to stocks held at the end of the year.

In addition to the usual questions about numbers engaged, turnover, stocks, etc. in the 1966 Census, there were questions about floor space for establishments in central shopping areas included in the sample.

Index Numbers of Retail Sales

An Index of Retail Sales was compiled for several years, before and during the Second World War, by the Bank of England in collaboration with trade associations. In 1947 a more detailed return was introduced and the work was taken over by the Board of Trade. For some years two series, one for Large Retailers and one for Independent Retailers, appeared monthly in the *Board of Trade Journal* and the *Monthly Digest of Statistics*, but they suffered from an inadequate and possibly biased sample of retailers and from the lack of a proper weighting system.

The 1950 Census of Distribution provided a much-needed sampling frame and a reliable weighting system, and indices of weekly values

Table 22.1 Turnover, stockturn and gross margin by kind of business and form of organization, 1957, 1961 and 1966
(Board of Trade Journal, 18 March 1970)

Kind of business and form of organization	Turnover 1957 (£ thous.)	Turnover 1961 (£ thous.)	Turnover 1966 (£ thous.)	Turnover divided by end-year stocks 1957 (Ratio)	1961 (Ratio)	1966 (Ratio)	Gross margin as a percentage of turnover 1957 (%)	1961 (%)	1966 (%)
TOTAL RETAIL TRADE	7,793,697	9,215,794	11,660,929	9.3	9.1	9.2	23.3	24.9	26.7
Co-operative societies	957,052	1,012,245	1,089,526	11.8	10.8	10.7	21.8	23.1	23.5
Large multiples	1,912,034	2,619,243	3,865,045	9.0	9.2	9.4	26.4	28.2	31.0
Independents	4,924,611	5,584,306	6,706,358	9.1	8.9	8.9	22.4	23.7	24.8
Organizations other than co-operative societies	6,836,645	8,203,549	10,571,403	9.0	9.0	9.1	23.6	25.1	27.1
Grocers and provision dealers	1,578,329	1,885,581	2,399,306	14.4	14.4	15.2	15.4	16.2	16.1
Large Multiples	456,865	638,944	1,033,223	13.3	12.8	14.4	17.6	18.0	17.8
Independents	1,121,464	1,246,637	1,366,083	14.9	15.5	15.8	14.6	15.2	14.8
Other food retailers	1,316,136	1,539,427	1,852,170	39.1	35.8	32.9	22.1	24.0	25.3
Large Multiples	290,971	420,313	622,859	24.9	22.2	20.4	24.1	26.3	28.7
Independents	1,025,164	1,119,114	1,229,311	46.8	46.4	48.0	21.5	23.2	23.6
Confectioners, tobacconists, newsagents	692,191	772,484	1,023,942	13.0	12.5	12.1	14.9	16.0	15.9
Large Multiples	74,109	80,032	135,152	10.5	10.3	10.2	17.7	21.3	21.2
Independents	618,082	692,452	888,791	13.4	12.8	12.4	14.6	15.3	15.0
Clothing and footwear shops	1,129,975	1,323,135	1,706,379	5.2	5.5	5.6	28.1	29.1	32.4
Large Multiples	463,968	628,496	891,411	6.5	6.7	6.9	27.7	29.1	33.4
Independents	666,006	694,640	814,968	4.6	4.7	4.7	28.3	29.1	31.4
Household goods shops	767,122	952,885	1,296,109	5.2	5.4	5.9	30.4	34.7	38.1
Large Multiples	152,724	229,939	371,841	5.7	7.0	7.8	39.5	51.0	53.8
Independents	614,398	722,945	924,268	5.1	5.0	5.4	28.1	29.5	31.8
Other non-food retailers	565,393	682,594	995,241	5.2	5.0	5.0	31.7	31.1	31.6
Large Multiples	146,956	195,316	288,542	5.9	5.8	5.5	33.7	31.3	33.5
Independents	418,437	487,279	706,699	5.0	4.7	4.9	31.0	31.1	30.9
General stores	787,500	1,047,443	1,298,254	8.9	8.6	8.9	30.9	32.0	37.1
Large Multiples	326,440	426,204	522,016	8.9	8.8	8.0	31.5	31.6	40.5
Independents	461,060	621,240	776,238	8.9	8.5	9.7	30.5	32.2	34.8

of sales were published for various kinds of business and for the three types of organization mentioned above; the panel of retailers was also extended to cover nearly all retail trade. These indices are still compiled in much the same way, being revised after each Census of Distribution, and based on the year of that Census. Those for broad groups are weighted arithmetic means of the component indices, with weights proportional to total turnover in the Census year.

Since 1961, indices of volume have been compiled, for broad groups only, as shown in Table 22.2. The reader will notice that they are seasonally adjusted. The tendency now is to regard seasonally adjusted figures as those of real significance, and to use more sophisticated methods of making these adjustments. There is, however, no satisfactory way of removing the effects of changes or expected changes in purchase tax, hire purchase regulations, etc., or of abnormally good or bad weather, and, however refined the methods may be, the seasonal factors are subject to considerable uncertainty.

Hire Purchase

Statistics of hire purchase are an important indicator of the country's economy. There are two main types of transaction: (i) hire-purchase agreements under which the customer hires the goods but does not own them until he has paid off the credit allowed him; (ii) credit-sale agreements under which the goods become his property from the outset. The former usually relate to large durable goods like cars and furniture; the latter to clothing, bicycles, etc. Credit may be obtained from finance houses, durable goods shops and department stores, or other instalment credit retailers such as mail order houses.

Various tables published monthly in *Trade and Industry* give figures of new credit extended, repayments and debt outstanding. The more detailed tables of new credit show percentage changes compared with a year earlier. The information is obtained from a sample of finance houses, stores and other sources of credit.

Stocks and Capital Expenditure

These two items are closely connected and may be discussed together as two aspects of capital formation. Quarterly returns of stocks and capital expenditure are rendered voluntarily by large numbers of firms in manufacturing industry, distribution and service trades, and the results are summarized in *Trade and Industry* and, in less detail, in the *Monthly Digest of Statistics*. The articles dealing with stocks show changes in the seasonally adjusted volume (i.e. value at base-year prices) held by various industry groups and kinds of distributors, and, in less detail, changes in book value of stocks.

(part of a table from the *Monthly Digest of Statistics*, June 1970)

Sales: weekly average 1966 = 100

	Seasonally adjusted — Volume						Unadjusted — Value	Total value of sales	Food shops			
	All kinds of business	Food shops	Non-food shops: Total	Clothing and footwear shops	Durable goods shops	Other non-food shops	All kinds of business: Value		Grocers	Butchers	Greengrocers, fruiterers	Fishmongers, poulterers
Sales in 1966—provisional estimates (£ million)	11,372							4,952	2,919	724	312	82
1966	100	100	100	100	100	100	100	100	100	100	100	100
1967	101	101	102	100	103	102	103	103	103	101	101	99
1968	104	102	105	104	108	104	109	108	109	105	99	99
1969	103	103	103	104	104	102	114	115	117	111	103	102
1968 4th quarter	104	104	104	105	105	104	126	117	119	114	93	103
1969 1st quarter	102	103	101	102	102	100	104	111	114	112	96	101
1969 2nd quarter	104	103	104	105	104	104	111	114	116	107	115	104
1969 3rd quarter	103	103	103	105	104	104	111	113	115	105	107	98
1969 4th quarter	103	103	103	103	105	105	132	123	125	119	96	103
1970 1st quarter	103	104	103	103	105	102	110	117	119	117	102	107
1970 2nd quarter	104	103	105	107	105	103	117	120	122	111	122	104
1970 3rd quarter	105	103	106	107	109	104	119	120	121	113	108	102
1969 July	102	102	102	105	101	102	111	112	114	100	118	97
1969 August	104	104	104	106	107	103	111	115	119	107	113	100
1969 September	103	103	103	104	105	102	110	111	113	108	95	98
1970 July	104	102	105	107	108	104	120	119	121	110	118	105
1970 August	105	103	107	109	110	105	119	121	124	112	114	104
1970 September	104	103	105	106	110	103	119	119	119	116	96	99

The articles on fixed capital expenditure also show base-year values, seasonally adjusted, by various industry groups and types of distributive and service trades. The total for manufacturing industry and for distributive and service trades (but not for industry groups) are broken down into New Building Work, Vehicles, and Plant and Machinery. Before 1963 these items were shown both at current values and at base-year values, not seasonally adjusted, so that it was possible to compile indices of costs of construction, vehicles, and plant and machinery. Although these figures are no longer published, they can be obtained from the Department of Trade and Industry on request.

Orders in Engineering Industries

One sensitive indicator of demand for capital equipment, and therefore of business confidence, is the state of the "order book" of the engineering industries, particularly heavy engineering. The monthly article in *Trade and Industry* on "Orders, Deliveries, Production and Exports in the Engineering Industries" contains two tables of index numbers of volume, i.e. value at base-date prices. One table gives orders on hand, net new orders, and deliveries for Engineering Industries, divided into export, home and total. The other shows indices of production, exports, and orders on hand for the Engineering and Electrical Goods Industries only. Every quarter there is a third table giving the actual value of deliveries of various products or product groups.

The Engineering Industries to which the first table relates include the engineering and electrical goods covered by the second, also locomotives and other railway equipment, heavy commercial vehicles and tractors, but exclude passenger cars, aircraft, and certain other items.

Inland Transport

Transport statistics have been described as "all bits and pieces," being compiled by various bodies with apparently little co-ordination. This impression is confirmed by the statement made by the British Road Federation that their *Basic Road Statistics* "is issued annually in an attempt to bring some order into the confusion which arises from the use of contradictory or conflicting road statistics."

However, the position is gradually improving. The Department of the Environment publishes two useful annual reports, (i) *Highway Statistics*, containing information on the nation's roads and road vehicles, and (ii) *Passenger Transport in Great Britain*, giving details of all kinds of passenger transport. A number of other reports are issued by British Rail, London Transport and other bodies.

References and suggestions for further reading

Trade and Industry
Monthly Digest of Statistics
Economic Trends
Reports on the Census of Distribution and other Services
SCHNEIDER, J. R. L. and SUICH, J. C., "The Census of Distribution for 1966," *Statistical News*, No 7 (1969)
MAKEPEACE, R. W., "Preparing for the Census of Distribution for 1971," *Statistical News*, No 16 (1972)

23

Overseas Trade and Balance of Payments

Importers and exporters are compelled by law to give particulars of all goods imported into, or exported from, the United Kingdom, including the value and, where required, quantity and the country of consignment (see below). Before 1970, there were different classifications for imports and exports, and importers had to classify their goods in two ways—for duty and for statistical purposes. Now both imports and exports follow the "Customs and Excise Tariff and Overseas Trade Classification."

Certain classes of goods are excluded from the detailed import and export statistics, such as personal effects and parcels brought or taken by passengers for private use, ships' stores and bunkers (coal and oil), and atomic energy materials. They are, however, allowed for in the Balance of Payments (see below), and more recently in the summary figures issued each month to the public. Exports are valued f.o.b. ("free on board"), which represents the cost to the purchaser abroad, including packing, inland and coastal transport in the United Kingdom and all other expenses up to the point where the goods are deposited on board the exporting vessel or aircraft.

Imports are valued c.i.f. ("cost, insurance and freight"), which broadly means the cost to the importer up to the port of entry. It does not include Customs duty or purchase tax. Again, however, the summary figures are adjusted to agree with the Balance of Payments, and the import figures totals are presented f.o.b.

Imports are classified as received from the country of consignment, i.e. the country from which they were originally dispatched to the United Kingdom, not necessarily the country of origin or manufacture. Indeed, the importer may not know the country of origin.

Exports are classified as dispatched to the country of final consignment, not necessarily the country of unshipment or consumption.

Before 1970 a distinction was made between "Exports" and "Re-exports," but there is now no such distinction.

Trade Returns of the United Kingdom

The original documents are sent to the Statistical Office of Customs and Excise, where the data are tabulated and summarized. The results appear in the following publications—

(a) *Annual Statement of the Trade of the United Kingdom*, published in five volumes. Volume I is a summary volume, giving the total quantity (where appropriate) and value of imports and exports of each item classified. Volumes II and III repeat this information for imports and exports respectively, with details for the more important countries concerned (see Table 23.1). Volume IV takes every country in turn and gives particulars of the principal goods imported or exported. Volume V gives details of the principal commodities for each port (sea or air) of the United Kingdom.

(b) *Overseas Trade Statistics of the United Kingdom*, published monthly showing imports and exports of each item classified, with some country detail, for the current month and the year to date.

(c) *Report on Overseas Trade*, published monthly, summarizes the broad pattern of this country's overseas trade. An important feature is an analysis of the main commodity groups of imports and exports subdivided into various countries or groups of countries such as North America, the Sterling Area, and EEC (European Economic Community).

Trade and Industry contains monthly articles on United Kingdom trade, giving summary tables of imports and exports by area and by commodity class (but not both together), all seasonally adjusted; quarterly articles on World Trade and Production, giving summary figures for world trade and for imports and exports of OECD countries, all in US dollars and all seasonally adjusted; also an annual article on World Exports and Manufactures.

Unit Value and Volume Index Numbers

For many years the Board of Trade (and now the Department of Trade and Industry) has published index numbers of volume of imports and exports, sometimes accompanied by indices of import and export prices. The term "unit value" is now used officially in preference to "price" because the indices are compiled not from price quotations for specific products, but from average values obtained by dividing the total value of a commodity or heading (which may not be quite homogeneous) by the total quantity.

Table 23.1   UK imports of fabrics, 1968

(*Extract from the Annual Statement of Trade of the UK*, 1968, Vol II)

Headings and countries from which consigned	1968	
	Quantities	Value
65121. Fabrics excl narrow fabrics etc —wholly cotton, woven—grey (unbleached) etc—other woven of width—N/E 39½ in.—of plain weave, weighing—over 3 oz N/E 6 oz per sq yd etc—101 to 150—	Sq yards	Cwt	£
India	6,624,691	15,192	398,965
Pakistan	29,660,610	80,489	1,704,039
Malaysia	694,401	1,637	45,566
Hong Kong	2,151,233	5,899	166,410
Canada	5,434,879	11,340	376,207
Other Commonwealth Countries	143,470	295	7,796
Irish Republic	503,010	1,043	46,762
Total Commonwealth Countries and Irish Republic	45,212,294	115,895	2,745,745
Soviet Union	4,966,884	11,753	224,893
France	1,061,595	2,366	69,670
Portugal	1,446,917	3,418	105,304
Hungary	1,008,098	1,944	51,044
Czechoslovakia	828,033	1,816	45,168
Turkey	1,186,386	3,085	66,648
Egypt	653,320	1,727	37,836
China	15,698,895	34,091	750,873
Formosa	1,025,961	2,230	58,006
Republic of Korea	633,332	1,234	35,968
Other Foreign Countries	800,266	1,813	55,151
Total Foreign Countries	29,309,687	65,477	1,500,561
Total	74,521,981	181,372	4,246,306

The present series of Unit Value Indices was introduced in 1963, together with a new series of Volume Indices. They are published monthly in the same article in *Trade and Industry* and, at the time of writing, are both based on 1961 and compiled with 1961 weights, which means that their product does not necessarily agree with the index of total value. In other words it is officially recognized that one cannot have at the same time a volume index which reflects only changes in quantities, and a price index which reflects only changes in prices, whose product is equal to the index of total value (see Chapter 16).

In both Unit Value and Volume Indices it is necessary to exclude from the calculations items which are subject to erratic fluctuations in size, quality, etc., where a quantity or average-value figure might be misleading, but all commodity headings that appear to comprise reasonably homogeneous products are included.

An index of particular interest to government is known as the *Terms of Trade*. This is the ratio of the Export Unit Value Index to the Import Unit Value Index, expressed as a percentage. When this is greater than 100, the terms of trade are said to be favourable; when it is less than 100, they are adverse.

Balance of Payments

The balance of payments covers all the economic and financial transactions between residents of the United Kingdom and those of other countries. For many years these transactions were summarized in three accounts, the Current Account, the Long-Term Capital Account, and Monetary Movements. However, it became increasingly difficult to distinguish between "long-term capital" and "monetary movements," and this form of presentation was abandoned in 1970 in favour of a simpler and more intelligible approach.

The emphasis is now on the *current balance*, i.e. the difference between current income and expenditure, rather similar to profit or loss over the year, and the *total currency flow* into or out of the country, including capital movements as well as the current balance. The current balance is built up from estimates, but the total currency flow is known precisely from official records of currency transactions. Thus, in 1969 there was a total currency flow of £743 million into the country which enabled us to repay £699 million of official loans and add £44 million to reserves.

The current balance is derived from *visible trade*, i.e. imports and exports of goods, both valued f.o.b., and *invisibles*, which fall into three main groups—

(a) Services, such as shipping and civil aviation, banking and insurance, travel, and government expenditure abroad on armed forces, embassies, etc.

(b) Interest, Profits and Dividends, i.e. income from overseas investments. In the case of British firms operating abroad, all profits, not merely distributed profits, are included, since retained profits are regarded as income reinvested. They therefore appear twice, as a credit item in this group and as a debit item in Private Investment.

(c) Transfers, such as gifts, legacies and aid to under-developed countries.

The import and export totals are believed to be fairly accurate, but the balance of visible trade, being the difference between two large and almost equal quantities, is relatively less reliable. The main weaknesses in the invisibles are private transfers and "other services" (finance, royalties and films, postal services, etc.). Figures for civil aviation are based on quarterly returns from airlines, and those for shipping on censuses about every fourth year and sample inquiries in other years by the Chamber of Shipping. Estimates for travel are based on interviews of a stratified sample by the International Passenger Survey, and are believed to be fairly reliable.

In theory, the total flow of currency should be equal to the current balance together with *investment and other capital flows*, but since these items are subject to errors of estimation there is usually a substantial discrepancy called the *balancing item*. Table 23.2 shows how the total currency flow of £743 for 1969 was made up. The figures are subject to revision in later years.

Table 23.2   Summary of UK Balance of Payments, 1969

	£ million
Exports (f.o.b.)	7,061
Imports (f.o.b.)	7,202
Visible balance	−141
Government services and transfers	−458
Private services and transfers	+564
Interest, profits and dividends	
Private sector	+780
Public sector	−329
Invisible balance	+557
Current balance	+416
Investment and other capital flows	+ 48
Balancing item	+279
Total currency flow	+743

Detailed tables and notes are given in the annual "Pink Book" entitled *United Kingdom Balance of Payments*, and following preliminary estimates in a White Paper, quarterly estimates are published in Economic Trends with a detailed commentary. Estimates for individual quarters, particularly of the current balance, are less reliable than the annual figures.

The remarks made about reliability should not be taken as in any way a reflection on the competence of our Government statisticians. They are well aware of the limitations of their statistics and do the best they can with very imperfect data.

References and suggestions for further reading

Annual Statement of the Trade of the United Kingdom
Overseas Trade Statistics of the United Kingdom (monthly)
Report on Overseas Trade (monthly)
Trade and Industry
Economic Trends
United Kingdom Balance of Payments (annual)
ALLEN, R. G. D. and ELY, J. E., *International Trade Statistics* (Wiley and Sons)

24

National Income and Expenditure

Sources

The three main sources from which the national income statistics are derived are the Census of Production, the statistics of tax assessments, and the accounts of the Central Government. There are a large number of minor sources such as the statistics of retail sales, Customs and Excise statistics, Balance of Payments figures, accounts of local authorities and public corporations, and so on. Since the data have been collected for different purposes and on different bases, it is a formidable task to fit them together, and the result has been aptly described as "essentially a synthetic product."

As already mentioned in Chapter 17, quarterly estimates of national income and expenditure are published in *Economic Trends* in the form of a special article. They also appear without commentary in the *Monthly Digest of Statistics*. Annual figures covering a period of years are given in more detail in a Blue Book (formerly a White Paper) on National Income and Expenditure, compiled by the Central Statistical Office. It will be referred to throughout this chapter as "the Blue Book." It is generally preceded by a White Paper giving Preliminary Estimates, published about four months earlier.

Threefold Nature of National Income

National Income may be defined as a measure of the goods and services becoming available to the nation for consumption or investment. It may be studied from three aspects, viz. how it is produced, how it is distributed, and how it is spent. Now the only practicable way of measuring the value of all goods produced and services rendered is to ascertain the incomes that make up the cost of production, i.e. wages, salaries, etc. including company profits, whether distributed or not. National Product and National Income are therefore equal by definition and derived from the same data, although their components may be shown differently in the tables. It must be noted, however, that National Income is generally taken to mean the net national income after allowing for capital consumption (see page 264).

Again, since incomes are either spent or saved, National Income is equal to expenditure on consumption plus addition to wealth, i.e. to National Expenditure defined in such a way as to include investment. It can therefore be built up from estimates of expenditure and investment, but this time the data are different and the resulting figure of National Expenditure will probably differ from that of National Income or Product, owing to errors of estimation.

Whichever approach is used, the aggregate itself is of limited interest. To economists, business men and government, the individual items that compose it are far more important.

National Income, a Measure of Economic Activity

It should be noted that National Income statistics are confined to economic activities, i.e. activities pursued for money. Work done for oneself or voluntarily for others does not normally enter into National Income, not because it is irrelevant, but because it cannot be measured. For example, if a man pays to have his house painted, the value of the work is included; if he does it himself, it is not included. The statistical difficulties in estimating the value of unpaid work and services would be tremendous. The principal item would, of course, be unpaid domestic services. This is a good example of the problem that often arises in statistics, whether to choose a second-best concept for which good data are available or a better concept for which satisfactory data do not exist. Also excluded are payments which do not correspond to goods or services received, i.e. transfer payments (see page 268).

Some people would like to exclude goods and services which they regard as "unproductive," such as the activities of the Armed Forces, the drink trade or the football pools. The economist as such is not concerned with such distinctions; it is sufficient for him that there is a measurable demand for such services. Admittedly, there are anomalies in the concept of National Income; for example, waste and congestion may inflate the National Income and make it appear that the country is better off. A full discussion of the problem, however, is beyond the scope of this book.

Imputed Income

In addition to actual money incomes, National Income includes a certain amount of "imputed" income, notably the "rent" of owner-occupied houses and such forms of income in kind as can be identified, e.g. the equivalent of the food and clothing of the Armed Forces. There is a good economic argument for including the services of all kinds of durable goods, but it would be impracticable to attempt to value the services of such articles as furniture and motor-cars. The inclusion of imputed rent removes anomalies that might otherwise arise, e.g. if the National Income were reduced as a result of tenants buying the houses they already occupied.

Depreciation and Capital Consumption

As already mentioned, National Income is taken to mean net national income or product after allowing for depreciation, or capital consumption, i.e. the annual use of capital through wear and tear. Until 1953 depreciation was measured by the total of depreciation allowances for purposes of taxation, but this was far from ideal. These allowances are notoriously inadequate in times of rising costs, and are subject to arbitrary changes in the Budget. Thus the increased initial allowances for plant and machinery in 1949 meant a rise of £200 million in the total for depreciation and a corresponding fall in the National Income having no real significance.

For a time the attempt to measure real depreciation was abandoned, and from 1953 to 1955 no estimates of net national income were published. In 1956, however, estimates of capital consumption at current prices were given for the first time, and they are now a regular feature of the Blue Books. These estimates rest on somewhat arbitrary assumptions about the life of assets, the way in which they depreciate with time, and the adjustments required to express that depreciation in terms of current values.

It will be convenient to discuss Gross National Product rather than National Income, except in general terms. Each of the three aspects of Gross National Product will now be discussed in detail, beginning with that of production.

The Approach via Production

Gross Domestic Product at factor cost is the total income from economic activity in the United Kingdom, or the value of all work done and services rendered valued at factor cost, i.e. at the prices paid to the various "factors of production"—labour, capital, land, etc. It is gross in the sense that no provision is made for depreciation. The main contribution comes from industry and resembles the net output of industry as defined for the Census of Production, except that it omits purchased services such as advertising and insurance. An adjustment is also made for "stock appreciation," discussed below.

With the addition of net income from abroad, Gross Domestic Product becomes *Gross National Product.* That is to say, while the Domestic Product is what the United Kingdom produces, the National Product is the total income of residents in this country from production anywhere in the world.

The constituent items of Gross Domestic and National Product are shown in Table 24.1. Table 11 of the 1970 Blue Book, from which

Table 24.1   Gross National Product in 1969 by industry or other source

Industry or other source	£ million
1. Agriculture, forestry and fishing	1,197
2. Mining and quarrying	678
3. Manufacturing	13,346
4. Construction	2,559
5. Gas, electricity and water	1,364
6. Transport	2,348
7. Communication	930
8. Distributive trades	4,193
9. Insurance, banking and finance (including real estate)	1,250
10. Public administration and defence	2,391
11. Public health and educational services	1,987
12. Other services	5,073
13. Ownership of dwellings	1,991
14. *less* Stock appreciation	−815
15. Residual error	−342
16. Gross Domestic Product at factor cost	38,150
17. Net property income from abroad	451
18. Gross National Product	38,601

these figures are taken, gives a run of eleven years, but for reasons of space only one year is given here. The figures are subject to revision in later Blue Books. Items Nos 14 and 15 require some explanation.

(*a*) *Stock appreciation* (Item 14). An attempt is made each year to divide the change in the value of stocks into (i) the part due to the change in volume, i.e. in the quantity of stocks held, and (ii) the part due solely to changes in the prices at which stocks have been valued. Only the first part is regarded as contributing to the National Product. The second part, called "stock appreciation," is treated as a capital gain or loss and must be deducted or added as the case may be. In 1958 this item was negative and therefore appears as an addition in the Blue Books.

Stock appreciation is estimated by applying price indices to as many categories of stock as available data permit, making certain assumptions about accounting methods which may in fact be wide of the mark. The resulting estimates are particularly uncertain in periods of rapidly changing prices.

(b) *Residual error* (Item 15). As already mentioned, the estimates of Gross National Product from income and from expenditure are built up from largely independent data, and the differences are shown as "residual error." Purely for convenience of presentation the estimates from expenditure are taken as correct, and residual error imputed to those from incomes, but the latter estimates are not necessarily less accurate than the former.

The Approach via Incomes

The Gross National Product can also be presented to show the wages, salaries, rents, profits, etc., comprising the net output of industry and the value of other work and services. The bulk of this total is personal income, but as Table 24.2 shows, profits of companies make a large contribution.

Table 24.2   Gross National Product in 1969 by type of share
(Adapted from the 1970 Blue Book, Tables 1 and 19)

Share in the gross national product	£ million
1. Wages	13,915
2. Salaries	10,270
3. Pay in cash and kind of the Armed Forces	543
4. Employers' contributions—	
National Insurance and health	1,141
Other	1,305
5. Total income from employment	27,174
6. Professional persons	541
7. Farmers	676
8. Other sole traders and partnerships	1,792
9. Total income from self-employment	3,009
10. Gross trading profits of companies	4,948
11. Gross trading surplus of public corporations	1,461
12. Gross trading surplus of other enterprises	114
13. Rent	2,601
14. *less* Stock appreciation	−815
15. Residual error	−342
16. Gross Domestic Product at factor cost	38,150
17. Net property income from abroad	451
18. Gross National Product	38,601

Items 6 to 13 are all gross before allowing for depreciation. Several items of this table need comments and explanations—

(a) *Wages* cover the gross earnings of manual workers and shop assistants before deductions for national insurance and workers' pension funds. This item includes an allowance for payment in kind, e.g. board and lodging for farm workers and domestic servants, and other concealed emoluments.

(b) *Salaries* cover the earnings of administrative, technical and clerical workers, including policemen, firemen and commercial travellers. The figures for salaries are less reliable than those for wages, and up to 1949 the total salary bill was obtained mainly as a residual figure by deducting wages from total earnings of employees.

(c) *Pay of the Armed Forces* includes large concealed emoluments such as food and uniform, also retired pay and pensions (other than for death or disability due to war service) of former members.

(d) *Employers' contributions* to national insurance, superannuation funds, etc. are regarded as part of the employee's income deducted at the source.

(e) *Professional earnings* are the earnings, net after allowance for expenses, of professions where profits depend mainly on personal qualifications, excluding salaried employees.

(f) *Income from farming* represents income from agriculture, horticulture and direct retailing of farmers with holdings of one acre or more, but not rent from the ownership of land. It is estimated by subtracting rent, wages and other costs from the gross output of agriculture.

(g) *Profits of other sole traders and partnerships* cover all the profits assessable under Schedule D of people in business for themselves or in partnerships not already accounted for.

Interest and dividends are not shown in Table 24.2 because they are included in Items 10, 11 and 12; nor are pensions, because they are not payments for current services. These items would, however, be included in any table of personal incomes.

Personal income generally means the income of an individual, or of a married couple, as assessed for income-tax, i.e. it includes all earnings, dividends, family allowance, etc., but excludes expenses deducted for tax purposes.

Total personal income includes not only the total of personal incomes as defined above, but also the (as yet) undistributed profits of life assurance and pension funds, imputed income from house-ownership, and various other items such as post-war credits and accrued interest on National Savings Certificates. It also includes "current grants to persons," i.e. National Insurance benefits, family allowances and other grants from public authorities, and interest on the National Debt.

The last group are often called *transfer payments*, because they are transfers of income from one group of persons, e.g. tax-payers or rate-payers, to another group, not payments for goods produced or services rendered (unless maternity benefits and family allowances can be so described). It should be noted that some transfer payments, notably school milk and welfare foods, are not income as generally understood, and are not regarded as income for purposes of taxation, while others, such as pensions and family allowances, are subject to tax. None of them, however, is directly related to production. Furthermore, if all incomes are added together, these payments are counted twice, once in the tax-payer's income and once in that of the recipient. Total personal income therefore contains an element of duplication, and if the Gross National Product is derived from it, transfer payments must be deducted.

The Approach via Expenditure

The goods and services made available for use during any given year include both domestic product and imports, and they can be either consumed (or bought for consumption), saved in the form of capital investment or additions to stocks, or exported. Table 24.3

Table 24.3 Gross National Product in 1969 by type of expenditure
(Adapted from the 1970 Blue Book, Table 1)

Expenditure generating gross national product	£ million
1. Consumers' expenditure	28,618
2. Public authorities' current expenditure on goods and services	8,118
3. Gross domestic fixed capital formation	7,927
4. Value of physical increase in stocks and work in progress	294
5. Total domestic expenditure at market prices	44,957
6. Exports and property income from abroad	11,986
7. *less* Imports and property income paid abroad	−11,318
8. *less* Taxes on expenditure	−7,868
9. Subsidies	844
10. Gross National Expenditure at factor cost = Gross National Product	38,601

shows how this brief analysis can be used to derive the Gross National Product from figures of expenditure, saving and foreign trade.

(*a*) *Consumers' expenditure* is expenditure on consumers' goods and services by persons and non-profit-making bodies, including the value of income in kind. In other words, it is not simply the amount of money spent, but the market value of the goods and services consumed. In the case of hire-purchase transactions, however, the whole of the purchase price is included.

(b) *Public authorities' expenditure* is current expenditure constituting a direct demand for goods and services, and excludes transfer payments and debt interest, capital expenditure, loans, etc.

(c) *Fixed capital formation* is expenditure on new buildings, plant and machinery, etc., whether for addition to, or replacement of, existing assets and excluding expenditure on repairs and maintenance, which is treated as a current cost of production.

(d) *The value of the physical increase in stocks and work in progress* is the increase in the value of stocks less "stock appreciation," and, like the latter, is subject to considerable error.

Item 5, *total domestic expenditure at market prices*, contains all Customs and Excise duty, purchase-tax, etc. included in the market price of the goods and services bought, and these items must be deducted to arrive at total domestic expenditure at factor cost. Similarly, any subsidies reducing the market price below factor cost must be added back.

The Blue Books give several tables showing details of consumers' expenditure, both at market prices and at factor cost, at current prices and also at fixed prices. Thus, in the 1970 Blue Book, expenditure is revalued at 1963 prices (see page 271).

The only way to estimate personal saving is to deduct personal expenditure from *personal disposable income*, i.e. personal income after tax and national insurance contributions. As residual figures, these estimates are subject to very large errors and subsequent corrections, which provoked *The Economist* to say, some years ago, "A separate chapter headed 'Where we went wrong last year' should be an annual feature of the White Paper."

Social Accounts

Most of the tables in the Blue Book are in the form of accounts, called *social accounts*, in which payments by, and receipts of, a group or *sector* of the economy are so defined that their totals must balance. It is customary to base these accounts on the following sectors—

The personal sector (including unincorporated business concerns)

Companies (private corporate trading bodies).

Public corporations (e.g. nationalized industries).

Central Government including national insurance funds.

Local authorities.

The rest of the world (a "sector" representing non-residents' transactions with the United Kingdom).

It may be observed that other countries do not necessarily divide their economies into the same sectors.

It is convenient also to distinguish between three types of economic activity, viz. production, consumption, and addition to wealth by capital formation. Each of these could in theory have its corresponding type of account, an Operating Account for production, an Income and Expenditure Account or Appropriation Account for consumption, and a Capital Account for addition to wealth, but in practice many of these accounts would not be appropriate. Table 24.4 shows the Combined Operating Account for Public Corporations for 1969.

Table 24.4　Combined Operating Account for Public Corporations, 1969
(Adapted from the 1970 Blue Book, Table 32)

Receipts	£ million	Payments	£ million
Sales—		Wages, salaries, etc.	2,761
(a) Revenue Sales	6,993	Purchases of goods and services	3,116
(b) Sales to own capital account	346	Plus Decrease in value of stocks and work in progress	28
Subsidies	160	Taxes on expenditure—	
		Rates	110
		Motor vehicles and catering licences, etc.	10
		Trading surplus and rent before providing for depreciation and stock appreciation	1,514
Total	7,499	Total	7,499

A distinction can also be drawn between types of transaction, e.g. payments for goods and services, and transfer payments.

For further details the reader should study the Blue Books, and in particular the admirable introduction to the 1952 Blue Book.

"Real" National Income

The chief difficulty in studying movements in the National Income or its components is that they are expressed in terms of money, the "value," or purchasing power, of which is liable to change from year to year. Attempts have therefore been made to measure the "real" National Income or product, i.e. the volume of goods produced and services rendered. There are two intelligible ways of doing this. One is to extend the methods and information used in compiling the Index of Industrial Production to cover agriculture, services, and other forms of economic activity, giving the results in the form of index numbers. The other is to revalue the goods and services comprising the real product in terms of prices and values current in a particular year.

Both methods are used in the Blue Books. For example, the 1970 Blue Book contains one table entitled "Expenditure and Output at 1963 prices" and another entitled "Index numbers of output at constant factor cost (1963 = 100)." The first table is derived from estimates of expenditure on goods and services, valued at 1963 prices. The second is based on the Index of Industrial Production and other statistics relating to volume of output.

Either method presents difficulties, particularly in dealing with services such as distribution and public administration. In the index number method, for example, insurance and medical services are measured by the number of beneficiaries served. The chief difficulty in the expenditure method is to find appropriate price indices for deflating current values. It is commonly assumed that the purchasing power of money is simply the reciprocal of the Index of Retail Prices, and Parliamentary questions and answers regarding the value of the £ are usually based on this assumption. In fact, however, this index only covers the field of retail transactions. For the remainder of the National Expenditure, e.g. public services and investment, reliable price indices do not exist. Moreover, the compilers stress that many estimates are extremely doubtful and that in many cases there are changes in quality that cannot be taken into account. It should be noted that the term "real National Income" is never used in the Blue Books.

Assuming that the real National Income of a country can be measured with reasonable accuracy, a rough indication of the average standard of living at different periods can be obtained by dividing the real National Income by the population. It must be remembered, however, that habits of life change considerably over long periods, and the age-composition of the population should also be taken into account.

International comparisons by means of real national income or product per head are even more difficult. Certainly there are official rates of exchange, but they do not always represent ratios of purchasing power. Comparison of British and American National Income per head will give different results before and after the devaluation of the pound in 1967. Two countries may differ in their treatment of transfer payments, the Armed Forces, domestic services, and many other matters. Moreover, in some countries much work may be done by the family, e.g. on the land, or by forced labour. Considerable progress has been made by the United Nations and the

OECD towards uniformity of concepts, but differences between highly-developed and backward countries seem to present insoluble difficulties.

Suggestions for further reading

Blue Books on *National Income and Expenditure*
National Income Statistics, Sources and Methods
United Nations Statistical Office, *Statistics of National Income and Expenditure*
United Nations Statistical Office, *Yearbook of National Accounts Statistics*
A Standardized System of National Accounts (OECD)
EDEY, H. C., PEACOCK, A. T., and COOPER, R. A., *National Income and Social Accounting* (Hutchinson)
STONE, R. and G., *National Income and Expenditure* (Bowes and Bowes)

Appendix 1

Automatic Data-Processing Techniques for Statistical Computation[1]

THE advantages of using automatic equipment for the processing of large volumes of statistical data and for the rapid production of results have long been appreciated. In fact the earliest forms of punched-card equipment were developed for that very purpose, for the rapid analysis of returns of the American census of 1888. More recent examples on modern equipment were those of the Pakistan decennial censuses, of the non-agricultural labour force and of housing and cottage industry, in 1961. For present purposes, examples of ICL data-processing equipment have been selected; the range of such equipment is very wide, from small and relatively inexpensive punched-card machines to huge electronic computers. Somewhere within the ICL range will be found the appropriate equipment for almost any kind of statistical work.

The two principal types of automatic data-processing equipment are those mentioned above: punched-card equipment and electronic computers, and, although there is some overlap (for example, punched cards are used as input media for computers), it is proposed to deal separately with these. Much of the work once carried out on punched-card equipment is now performed on a computer; however, the equipment described in the following section is still used in various departments.

Punched-Card Data-Processing Equipment
The functions of these machines, though many and varied, fall naturally into four main groups, those of translation, arrangement, calculation and presentation.

[1] Prepared by D. Walley BSc. The first part is based on the previous appendix by J. W. Parker BSc, PhD, LIB.

1. *Translation functions.* Statistical data may be contained in a variety of source documents, e.g. census returns, completed questionnaires, schedules, meter records—even the rough jottings of the statistician himself. Since no machine has yet been invented to read the entries directly from these source documents, they must first be translated into a language which the machine can read and understand, namely, holes punched in significant locations in special cards.

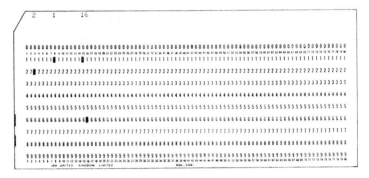

Fig. A.1   Punched Card

The cards are divided horizontally into "columns," the popular standard size of card being 80 columns. Groups of such columns taken together are called "fields." The cards are also divided vertically into "positions," representing the numerical values "0" to "9," reading downwards. The significant location of any hole punched in the card is defined by its column number and its position in that column. The card illustrated (Fig. A.1) records the following information—

Identification No 62479	Sex (code)
Date of Birth 22 April '09	Marital Status (code)—
County (code) 29	S. 1, M. 2, W. 3, D. 4
Town (code) 136	Occupation (code) 014
	No and sex of children (M1, F2)

The foregoing might represent the first few entries on, say, a census return, the remaining data from which would be punched in as many as necessary of the unpunched columns on the card, divided into appropriate fields. In this particular example one such card would be punched for the head of the household, but in other applications a card for each person might be required.

Punching may be effected rapidly by the use of an *Automatic Key Punch* wherein the card is automatically positioned and the depression of the appropriate key causes a hole to be cut in the required position and the carriage to move forward, normally one column, though "skipping" may be carried out. As noted above, the value of each hole is completely determined by its position. Facilities are also available for the automatic recording of common information in any number of cards and for the automatic reproduction of holes punched in selected columns of one card into the same or any other columns of another, with change of order if required.

The accuracy of translation of the data is "verified" by a second operator, using the source documents and a different machine (the *Verifier*), repeating the original punching operation on the cards. If any disagreement is detected the card is investigated and if necessary repunched.

Other punching equipment includes *Reproducing Punches* for reproduction of the information into other cards; *Gang Summary Punches* for use in conjunction with the *Tabulator* (later described) to produce, automatically, summary cards or new balance cards recording the results of calculations carried out on the tabulator.

2. *Arrangement functions*, whereby the cards punched as above, and the data contained in them, can be arranged in various ways to permit subsequent operations, and can also be counted. The principal machines used for this purpose are—

(*a*) *Sorters*, by using which the cards may be arranged in whatever order is necessary for further processing. Thereafter, by further sorting, the cards may be rearranged in a different order to satisfy other requirements.

(*b*) *Collators* for merging in some desired order two or more separate packs of cards to produce a single pack, or performing the reverse operation of sorting a merged pack into its original constituent packs.

One machine in this group which is of particular interest to statisticians is the *Counter Sorter* which is the simplest machine available for rapid statistical analysis. This machine can sort cards into separate packs, counting and registering the number in each pack and the total number in all packs at speeds of about 24,000 cards an hour. If necessary it can simply count and record without disturbing the sequence of cards in the original pack. By its use, for example, the numbers each of single, married, widowed and divorced persons in a population of 100,000, for whom cards had been punched as in Fig. 44, could be determined in just over four hours with, for control purposes, the total number of cards passed through the Sorter.

3. *Calculation functions*, whereby basic arithmetical operations are performed on numerical information punched in the cards. Among the equipment which performs this function is the *Tabulator*, which can be used for carrying out many statistical computations, such as the summation of squares and of products, e.g. in regression analysis. Other (presentation) functions of the tabulator will later be considered.

There remains to be mentioned the important range of ICT "plugged program" equipment, all designed to exercise calculating functions. These machines were developed to provide extra computation facilities within the normal punched-card installation and occupy an intermediate position between the electro-mechanical punched-card machines and the computers later to be considered. By the use of electronics their calculating speeds were increased and they can, during any single passage of cards through them, carry out for each card a comparatively short series of calculations as prescribed in their plugged programs, punching the result of such calculations into the card itself. Many quite complex statistical calculations have been and are being carried out on these machines.

4. *Presentation functions*, whereby the results of all that has gone before—translation, arrangement and calculation—are presented in usable form. One example of equipment in which presentation functions are exercised has already been mentioned—the *Counter Sorter*, on which the results of counting are built up unit by unit in the various counters, to be read off and written down at the end of the operation. More useful is the *Tabulator* (also mentioned already), which can print on plain paper, or in correct positions on printed forms, information derived from cards, either card by card (Listing) or as accumulated summaries of calculations—usually additions and subtractions—on a group of cards (Tabulating). For presenting in punched-card form the results of the tabulator's listing or tabulating operations, the *Gang Summary Punch* mentioned on page 275 is used. Another machine used for presentation may be mentioned— the *Interpreter*, which is used to print "in clear," along the top edge of each card, what is punched in the card. This machine is of use in the preparation of pre-punched cards stored in "pulling files" and pulled therefrom when required for use.

So ends our survey of the main functions of punched-card data-processing equipment. For many statistical purposes, especially when the volume of data to be processed is not large, such equipment will be found entirely adequate, and will save the statistician many weary hours of purely routine work and reduce the probability of errors due to human fallibility and fatigue.

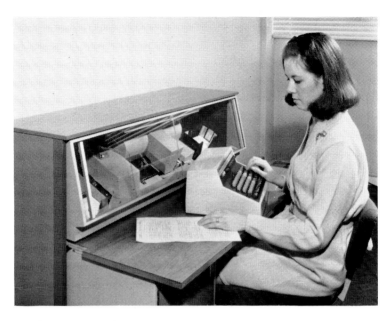

Plate 1 *Top* Automatic Key Punch
Bottom Card Counting Sorter

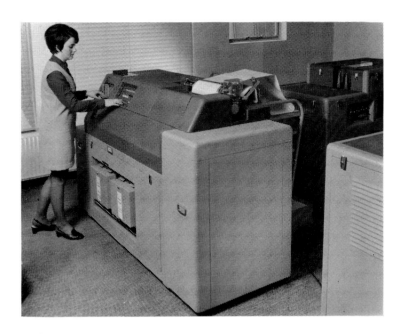

Plate 2 *Top* ICL Tabulator
Bottom ICL 1905 Computer System

Electronic Computers

The use of electronic digital computers for performing large-scale analyses has developed dramatically over recent years, and their application to the field of statistics is no exception. It is impossible within this appendix to give more than a very brief introduction to the ways in which a computer can help statisticians and to suggest a few references for further introductory reading.

A computer may, for our purposes, be divided into four sections—

Input for reading information from punched cards and paper tape, etc. For example, the cards described above could be read prior to calculation.

Output for presenting results of calculations in a readable form, or for providing information for future calculations.

Store for holding data, intermediate results and also the instructions (program) for the calculation required.

Arithmetic and *Control Units* for actually performing the required arithmetic and organizational operations as instructed by the programmer, typically at speeds of about a million instructions per second.

Each of these sections is designed to perform very quickly and accurately, thus allowing the computer to calculate reliably at very high speed, and reducing a week's calculation to a few seconds.

The computer has built into it a basic set of instructions which it is able to obey, and any calculation must eventually be specified in terms of these instructions. This would be a very tedious task if it were always left to the programmer and, because of this, programs called *compilers* have been written to help. They essentially accept problems written in an "algebraic" or "plain English" form and then translate these into the required basic form. We can therefore write such instructions as

AVE = TOTAL/NUMBER

which is understandable without any knowledge of computers.

It is not usual for a statistician to understand the basic machine instructions, but rather to write programs in a language to be accepted by a suitable compiler, e.g. FORTRAN, ALGOL, EMA, COBOL, PL/I, etc. He will find on learning one of these powerful computer languages that he is able to specify complicated calculations with ease, particularly as facilities exist for standard mathematical procedures, e.g. square-rooting, taking logarithms, etc. A simple program in FORTRAN to find the mean, variance and standard deviation of a set of fifty items is included here with brief comments.

20 FORMAT (F10.0)	Define input fields
21 FORMAT (1X, 3F10.5)	Define output fields
SUM = 0.0	Zero sum
SUMSQ = 0.0	Zero sum of squares
DO 1I = 1, 50	Repeat next 3 instructions 50 times
READ (5, 20)X	Input one number
SUM = SUM + X	Accumulate sum
1 SUMSQ = SUMSQ + X * X	Accumulate sum of squares
XMEAN = SUM/50.0	Form mean
VAR = (SUMSQ − SUM − SUM*50.0)/50.0	Form variance
STD = SQRT (VAR)	Form standard deviation
WRITE (6, 21) XMEAN, VAR, STD	Output results
STOP	Terminate program
END	Terminate compilation

This should be followed by the fifty data cards.

It can be seen that most of this is understandable without prior knowledge of FORTRAN. Looping instructions, such as "DO 1 I = 1, 50" are particularly important as otherwise the following three instructions would have to be written down fifty times to obtain the desired result.

Very often programs already exist for standard analyses which a statistician may require, either supplied by the computer manufacturer or obtained from a colleague, but it is important here to note that programs are not universally accepted by all computers—there may not, for example, be a suitable compiler available. If it is possible to use an existing program, all that is then required is for the data to be presented in the form that the program specification dictates.

A similar procedure is to use a program from a statistical package, for example, the BMD package (biomedical programs) which is available on most modern computers. Here one is required to present data to the package together with other information to define the type of analysis required. This package processes the data and provides the results of the analysis together with other detailed statistical information about the data.

The main advantages of using an existing program are the saving of effort and the fact that it should not contain errors. The best alternative is to choose a language where it is possible to specify the calculation with a minimum of effort. Specialist languages have been designed for particular areas of interest and ASCOP is such a language for statistics. Using this, one can specify concisely most statistical calculations involving large amounts of data with more flexibility than would usually be available. Some knowledge of both statistics and computing is required in order to use such a language.

More information regarding the languages mentioned here may be found in manuals and in many explanatory books, but the reader may prefer to continue in this more general vein, and the following books are suggested for this purpose.

Suggestions for further reading

General

JACOBOWITZ, H.   *Electronic Computers Made Simple* (Allen)
National Computing Centre   *Basic Training in Systems Analysis* (Pitman)
EMERY, G.   *Electronic Data Processing* (Pitman)
MILTON/NELDER   *Statistical Computation* (Academic Press)

Specific Systems

McCRACKEN, D. D.   *A Guide to FORTRAN IV Programming* (Wiley)
McCRACKEN, D. D.   *A Guide to COBOL Programming* (Wiley)
McCRACKEN, D. D.   *A Guide to ALGOL Programming* (Wiley)
NICOL, K.   *Elementary Programming and ALGOL* (McGraw-Hill)
SANDERSON, P. C.   *Computer Languages* (Butterworth)
COOPER, B. E.   *ASCOP User Manual* (National Computing Centre)
DIXON, W. J.   *BMD Biomedical Computer Programs* (UCP)

Appendix 2

Suggestions for General Reading

MANY references and suggestions for further reading have already been given, particularly in the later chapters, but a little more general guidance may be helpful. There are many (far too many) books on statistics, and the author has limited himself to a few with which he is acquainted and which he considers suitable for readers of this work.

Some readers may find this book a little concise and may wish to read other books covering similar ground but in a more discursive manner. Two excellent books for this purpose are Croxton, Cowden and Klein's *Applied General Statistics* (Pitman), and Wallis and Roberts's *Statistics: A New Approach* (Methuen). These are both large and fairly expensive books, going somewhat farther than this book into the theory of statistics, but explaining everything very clearly and in great detail. Croxton, Cowden and Klein is particularly useful for its chapters on time series and index numbers. Wallis and Roberts is valuable for its hundreds of examples and exercises.

A smaller and remarkably cheap book, of more general interest that its title would suggest, is Bradford Hill's *Principles of Medical Statistics* (The Lancet Ltd). In a very simple and lucid manner it deals with the collection and presentation of data, averages and dispersion, the elements of sampling and statistical tests, and the principles of experimentation, carefully pointing out the difficulties and fallacies encountered in these operations. There are also excellent chapters on standardized death rates and other indices, and life tables.

Ilersic's *Statistics* (H.F.L. (Publishers) Ltd) covers much the same ground as this book but is aimed at the less mathematical student. It contains excellent chapters on sampling methods and surveys, vital statistics (including a detailed account of the construction of life tables), and economic statistics.

Another book which has already been mentioned in Chapter 17, and which provides a very readable introduction to official economic statistics, is Lewes's *Statistics of the British Economy* (Unwin). It is to be hoped that both Ilersic and Lewes keep their books up to date.

The student wishing to pursue the mathematical theory underlying statistical techniques will find an admirably clear exposition of the subject in Hoel's *Elementary Statistics* and in his more advanced *Introduction to Mathematical Statistics*, both published by Chapman and Hall.

Answers to Exercises

For reasons of space, and also because there is often no single correct answer, no answers are given to exercises involving charts, tables or general discussion. (Even with those given below, there may sometimes be differences of opinion.) On the other hand, hints for solution and comments on the questions themselves are given when required. Answers are rounded off as appropriate.

Chapter 6

6.1   3·03 per cent; 0·182 cubic feet.

6.2   1·4648±0·0029.

6.3   Estimate 4·023, minimum and maximum possible values 2·935 and 5·278.

6.4   It is not quite clear what is expected, short of doing nine multiplication sums (without a machine, in the examination!) and three additions. The estimated, minimum and maximum values for the total are 2,174,360 tons, 2,165,682 tons and 2,183,052 tons.

6.5   Actual errors are −£383, etc., but to readers of the rounded figures the possible errors will be ±£500 on each item (±£3,000 on the total if it has been obtained by adding the rounded figures).

6.6   31·15 m.p.g.; 28·3 to 34·6 m.p.g.

Chapter 8

8.1(d)   ±5 million (most errors will be much less).

8.2   Period is 6 years, from peaks and troughs. Take 6-point moving totals, then add pairs and divide by 12.

8.3   Seasonally corrected figures: January £1,250, February £1,222, March £1,211.

8.4   Average seasonal movements: Spring, +10; Summer, −24; Autumn, −11; Winter, +25. (The question does not state the unit.) The first model fits well.

8.5 Be careful: there are five deviations for each of Quarters III and IV, but only four for I and II. Average them before adjusting to total zero. The average seasonal effects are: I, −£672; II, +£234; III, +513; IV, −£75.

8.6(c) Probably about 265 million square yards. (*N.B.* The actual figure turned out to be 257).

8.7 About 1,700,000 and 3,200,000. (Actual figures were 1,420,000 and 3,231,000. The alternative model would clearly be more suitable in this case. It would give 1,430,000 and 3,220,000.)

Chapter 9

9.1 State what upper limit you assume for the group "over 8 tons."

9.2 A poor question. There are too few employees to draw a reliable frequency curve, and the distribution is not symmetrical as the question implies.

9.5 Check your answers from the original data, remembering that "1" means "between $\frac{1}{2}$ and $1\frac{1}{2}$."

Chapter 10

10.1(b) The result will vary slightly according to your grouping. Compare it with the average of the original data.

10.2 93·84 lb per sq inch.

10.3 The data are somewhat peculiar: seven items of 168 and only one item between 152 and 168! Who's been cooking the records? Your answers will depend somewhat on your grouping, but compare them with the mean (160), median (168) and mode (168) of the original data.

10.4 Arithmetic means (in minutes): *A*, 15·76; *B*, 13·40. Medians: *A*, 13; *B*, 15. To explain this anomaly, try drawing cumulative frequency polygons.

10.5 28 mph; 19 per cent.

10.6 Mean 25·44 tons, median 25·0 tons, mode 20·9 tons, but the mode does not make sense with only 12 discrete items. If we assume that these are 12 random items from a large population, we can estimate the mode as 24.12 from the formula

Mode = Mean −3 (Mean − Median)

But does it really mean anything?

10.7(i) 126.

(ii) 110.

10.8 Arithmetic mean, 9·375; geometric mean, 8·17; harmonic mean, 6·99. (Some arithmetic for 36 minutes!)

Chapter 11

11.1 Impossible to give precise estimates, but medians about 48·5 in 1962 and 50·3 in 1966; quartile deviations nearly 8 and 9 respectively.

11.2 Median 99, quartiles 90 and 108.

11.3　Approximately 4′ 10·9″, 5′ 0·0″, 5′ 0·9″, 5′ 1·6″, 5′ 2·3″, 5′ 3·0″, 5′ 3·7″, 5′ 4·5″, 5′ 5·7″. The decimal figures are uncertain.

11.4　Very nearly 6 lb, measured from either mean or median.

11.5　Mean about 2·2 tons, standard deviation about 2·1 tons. More precise estimates depend on assumption made about "over 8" group.

11.6　£3·9 or, with Sheppard's correction, £3·8.

11.7　Mean 82·080, standard deviation 3·673.

11.8　Coefficients of variation: English 47%, Arithmetic 40%.

Chapter 12

12.1　(a) $\frac{81}{256}$ (b) $\frac{3}{8}$.

12.2　81 per cent.

12.3　(a) Mean 171·94 lb, standard deviation (with Sheppard's correction) 29·82 lb.
　　　(b) 210 lb.

12.4　(a) 9·1 per cent.
　　　(b) 2·9 per cent (new mean 26·75 lb).

12.5　(b) 0·942 (Poisson approximation gives 0·938).
　　　(c) 10.

12.6　$50e^{-1·8} = 8·3$.

12.7　(a) e^{-4}.
　　　(b) $\frac{8}{3}e^{-1}$.
　　　(c) $6e^{-6}$.

Chapter 13

13.2　Highly significant; the campaign was successful (but what a laborious question without a machine!).

13.3　Primer A is almost certainly superior ($P < 0·01$).

13.4　$(41 \pm 3·0)$ per cent, or 38 per cent and 44 per cent.
　　　Difference of 11 per cent is highly significant.

13.5　(a) 34 per cent.
　　　(b) $(34 \pm 4·15)$ per cent, or say 30 per cent and 38 per cent.
　　　(c) About 2,150.

13.6　(a) 9,500.
　　　(b) Stratify to remove bias.

13.7　The first part is stupid. The seedsman expects 40 seeds to germinate and 45 do. Of course he is justified.
　　　The stock is almost certainly sub-standard. The two results are not consistent.

13.8　0·73 to 3·27. No evidence of change: difference between sample means is 1, standard error of difference is 0·95.

13.9　7, using a double-sided test.
　　　3 hrs 12 mins; 3 hrs 2 mins and 3 hrs 22 mins.
　　　Yes: the confidence interval does not contain the old time.

13.10　Yes: $t = 2·32$ for 9 df, $P < 0·05$.

13.11　Yes: there are 9 positive differences and one negative one, giving $P = 11/512$. Using the t-test, $t = 3·57$ for 9 df, $P < 0·01$.

Chapter 14

14.2 (a) Any estimate between 20 and 21 knots is reasonable.

(b) Probably about 25 knots, but the relationship is not linear and the shape of the curve outside the range 6,000–30,000 tons is very uncertain.

14.3 $y = 1 \cdot 1786x - 4 \cdot 5491$.

14.4 0·8. But fancy only doing this for five pairs of observations!

14.5 (a) 0·88.

14.6 $y = 8 \cdot 24x + 5 \cdot 43$; 13·67; 23·56; 26·03.

14.7 $y = -1 \cdot 3068x + 4 \cdot 5632$.

Death rate (85 and over) = 36,580 (mean air temperature) $-1 \cdot 3068$.

It would be unsafe to use this formula for temperatures outside the range given.

14.8 Spearman's rank correlation coefficient is 0·82, but it is impossible to assess its significance with only seven items.

14.9 $-0 \cdot 14$. Not significant.

Chapter 15

15.1 1·4 per cent.

15.3 (b) 2·28 per cent.

15.7 Assume particles randomly distributed and use Poisson distribution with parameter 1.

Yes: actual and expected frequencies agree well.

15.8 Process out of control, i.e. quality is variable.

15.9 (a) 3·4 per cent.

(b) Inner limit 9·4, outer limit 12·1.

(c) No. The mean proportion would be the same for smaller samples (presumably "batches" is a mistake for "samples"), but the standard error of the proportion would be increased.

Chapter 16

16.1 Price index for 1960, 120; for 1962, 85.

16.2 (a) 102·69.

(b) 102·78.

16.3 Your answer will probably be Laspeyres, 123, or Paasche, 118.

16.4 (a) $-0 \cdot 6$ per cent (further precision on this item would not be justified).

(b) $+2 \cdot 27$ per cent.

(c) $+2 \cdot 55$ per cent.

16.5 102·0 (the base-weighted index is, in fact, 106·9).

16.6 (a) Divide Col (1) by Col (3).

16.7 Possibilities include value at 1962, 1963 or 1964 prices, simple arithmetic mean of production relatives, etc.

16.8 (a) $\Sigma q_{68}p_{58} = 3{,}954$, $\Sigma q_{58}p_{68} = 5{,}969$.

(b) (i) 269; 178.

(ii) 37·3; 56·3.

Chapter 18

18.1   (*b*) (i) 14 per 1,000.
(ii) 11 per 1,000.

18.2   The answer will depend on the standard population chosen, but groups "under 20" and "60 and over" must be omitted, as they have no means of comparison.

18.3   Actual, 11·60 days; expected, 10·99 days.

18.4   Crude death rate, 13·2 per 1,000; standardized death rate, 13·5 per 1,000.

18.5   134.

18.6   6,787 school places needed; 31,024 jobs required; 71 per cent.

Index